High-Temperature Superconductivity
An Introduction

Gerald Burns
*IBM Thomas J. Watson Research Center
Yorktown Heights, New York*

ACADEMIC PRESS, INC.
Harcourt Brace Jovanovich, Publishers

Boston   San Diego   New York
London   Sydney   Tokyo   Toronto

ACADEMIC PRESS, INC.
1250 Sixth Avenue, San Diego, CA 92101

United Kingdom Edition published by
ACADEMIC PRESS LIMITED
24–28 Oval Road, London NW1 7DX

Library of Congress Cataloging-in-Publication Data:

Burns, Gerald, date
    High-temperature superconductivity : an introduction / Gerald
Burns.
        p.   cm.
    Includes bibliographical reference (p.   ) and index.
    ISBN 0-12-146090-8 (alk. paper)
    1. High-temperature superconductivity. I. Title.
QC611.98.H54B87   1992
537.6'23—dc20   91–32808
                                                                      CIP

Printed in the United States of America

91 92 93 94   9 8 7 6 5 4 3 2 1

Dedication

This book is dedicated to Frank H. Dacol.

Friend, colleague, mentor.
He has always been there,
and has helped in every way.

CONTENTS

Preface

This book is written for those with a technical background who want to learn more about the exciting, new (post-1986) field called high-temperature superconductivity. The high-T_c field, as it is often called, has caused a worldwide revolution in solid-state physics, with thousands of people working in this area. The reason is that superconductivity now occurs at temperatures above the boiling point of liquid nitrogen, which, before 1986, was thought to be absolutely impossible. These superconductors already are finding some technological use.

Along with this revolution, excitement, and vast outpouring of work, conflicting results have been reported. The high-T_c materials have complicated chemistry and some conflicts have resulted from studying (unknowingly) slightly different materials. This aspect of the field is now settling down with more careful, more consistent experimental results being published. It is possibly fair to say that there is less convergence on the theoretical side. The ideas still range on one side to BCS-like, with electron-phonon coupling plus perhaps some "booster" to increase T_c value. The other side almost seems unbounded. However, experiments are beginning to restrict the possibilities.

I found writing this book more difficult than my other books. In this field, the important issues rapidly change; even experimental results can change. This book is prejudiced, as I believe any high-T_c book must be. The prejudgements arise in deciding what topics, sets of experimental results, and then what sorts of theoretical ideas to present. Most fields in physics, in which books are written, are mature enough to severely reduce (although not eliminate) the prejudgement problems; the high-T_c field is not. My prejudgements become particularly apparent in Section 5-6. Nevertheless, I felt that an introductory text would be useful for students who have heard, and are enthusiastic, about this field. To help overcome the prejudgement problems, each chapter has Notes, listed by topic in the back of the book. The references are to review articles, a few of the more basic papers, and some of the recent papers referring to topics discussed explicitly in the text (including some 1991 references to help the student). More extensive references to the original literature would be out of place in a book like this (and would very soon be useless, since the field is moving too rapidly). Even papers in the original literature run into this problem, resulting in omissions and prejudgements of other papers. By using the Citation Index for the references in the Notes and those in the figure captions, more recent work may be found. More importantly, review articles and conference proceedings keep appearing; these should be consulted for recent developments.

In deciding on topics, experimental results, and theoretical ideas, I have tried to keep to these thoughts: Do they have enduring quality? That is, might they still be true and possibly even important in a few years? Second, are they pedagogically simple enough to explain to students taking their first solid state physics course? The rapid developments in this field make these decisions difficult. In the course of writing this book, I have become impressed by the fact that many of the issues in high-T_c superconductors have been met earlier in studies of conventional superconductors.* This is the third guide to the chosen topics, namely, the similarities of conventional and high-T_c superconductors. These thoughts have persuaded me to say more about strong-coupled BCS than others might do at this time. I hope my choice of topics and approach will be of use to students and to people working in the field.

Mostly, an experimental point of view is taken in the book. I do not feel that many of the interesting and exciting theories of both the normal state and superconducting state of the high-T_c cuprates meet the guidelines. It is just not an easy task to describe the resonating-valence-bond theory applied to high-T_c materials, to discuss slave bosons in order to understand van Hove singularities, or to discuss the many other interesting theoretical approaches. Also, I am less sure which will be important in a few years.

I have been accused by one reviewer of the manuscript of taking a conservative point of view in my discussion of the field. I have not given enough space to other exotic mechanisms, and some of the real peculiarities of the normal state. The unusual c axis conduction, for some of the high-T_c materials, probably can not be understood in terms of conventional Fermi-liquid theory. These comments are all true. Thus, I urge readers to see the forthcoming book by P. W. Anderson, which will certainly look at high-T_c materials from a point of view that is different from that presented here. I also suggest that readers look at the discussion in the June, 1991

---

*Conventional superconductors are defined as noncuprate ones. Of course this includes all pre-1986 superconductors as well as interesting ones discovered since 1986. See Section 2-7.

Physics Today by P. W. Anderson and J. R. Schrieffer. It is a real pleasure to listen in on such discussions. Part of the problem is that only the BCS model is well developed after decades of research. Most of the other models of superconductivity, certainly some of the recent ones, are not developed far enough to allow detailed comparison to most experiments.

This book certainly is not meant to be the last word on high-temperature superconductivity. It is merely meant to help beginners obtain an overview of this field, while they may get involved with some area(s) in detail. I do apologize for all of the topics that could have been included and for the inadequacies of presentation of those that have been included. I am painfully aware of both shortcomings: the lack of time and space are inadequate excuses, but the only ones that I have. I have not included anything about the revelations going on with the discovery of the buckminsterfullerene C_{60} spherical-like molecules. When intercalated with K, Rb, or Cs with approximate formulae K_3C_{60}, Rb_3C_{60}, or Cs_3C_{60} they appear to have superconducting transition temperatures of 18 K to ~30 K, respectively. There is intense research on these "buckyballs"

It has been said that authors do not finish books; they abandon them. I think this is largely true, at least for science texts. After a while, the "fine tuning" becomes too picky and one has some vague idea to "start afresh," but that is impractical. At that time, the book goes to the publisher. This book certainly is not finished, but abandoned. Each week or so, during a visit to the library, I find 10 new interesting papers with 20 references to be read and this effect shows no signs of stopping. Hence, I am giving myself a lovely present of abandoning this joyful, educational, but burdensome effort.

Gerald Burns
Yorktown Heights, N. Y.

Acknowledgements

First and foremost, I would like to thank Frank H. Dacol for the huge amount of support, help, and thoughtfulness he has provided that enabled me to finish this and most of my other projects.

The arduous task of typing, retyping, and retyping was skillfully done by Phyllis Helms and Stephanie Rees. Their pleasantness and speed helped me a great deal. Betty McCarthy, our text consultant, put the finishing touches to many things and her efforts are greatly appreciated. Evelyn Marino entered the hundreds and hundreds of copy editing corrections and I thank her a great deal.

It is a pleasure to acknowledge the countless number of discussions with my many colleagues at the IBM Yorktown Heights Laboratory. Hardly was there a lunchtime without an enjoyable high-T_c discussion. In a sense, this book is a bound collection of ink-stained, lunchtime napkins. Without the help of all of these people, this book would have been impossible to undertake, no less complete.

In particular, I would like to thank Andrew Burns (Ohio State Univ.), Tracie Burns (Yale), Leanardo Civale, G. V. Chandrashekhar, Rowan Dordick, Chris Feild, Matthew Fisher, Fred Holtzberg, Roger Koch, Lia Krusin-Elbaum, Dennis Newns, Carol Nichols (Cornell), Tom Penny, and Ted Schultz.

Some of this book was written with the help of a Senior von Humboldt Fellowship and as a guest of the Max-Planck Institute for Solid State Physics, Stuttgart, Germany. The help and hospitality of both of these institutions is very gratefully acknowledged.

Chapter 1

Introduction

This was the greatest disturbance in the history of Hellenes,
affecting also a large part of the non-Hellenic world, and indeed,
I might say, the whole of mankind.

Thucydides, "History of the Peloponnesian War"

Before 1986, the "record high" for a superconducting phase-transition temperature was $T_c \approx 23.2$ K for Nb_3Ge (1973), and it was widely felt that this T_c value could at best be improved by only a degree or two in some exotic metallic alloy. In fact, theories were developed that purported to show why this should be the case (Section 2-6c). Hence, superconductivity was considered a "mature" field, and the search for higher-temperature superconductors had been largely abandoned.

Then, in 1986, two European scientists made a remarkable discovery. They had taken the point of view that higher T_c values might be found in materials in which the Jahn-Teller effect could enhance the electron-phonon coupling parameter. After years of effort, in another example of the power of "small science," they discovered a new approach for obtaining superconducting materials with transition temperatures higher than had been thought possible. J. Georg Bednorz and K. Alex Müller, working at the IBM Research Laboratory in Zurich, Switzerland, found a material with T_c in excess of 30 K in a class of cuprates (Cu oxides). The material was La_2CuO_4 in which ions of Ba^{2+}, Sr^{2+}, or Ca^{2+} had been introduced to replace some of the La^{3+}. The Sr-doped material is usually written $(La_{2-x}Sr_x)CuO_4$.

Poorly conducting, brittle Cu oxides were hardly the type of material in which scientists expected to find high-temperature (high-T_c) superconductivity. It wasn't until late 1986 and early 1987, when their results were verified in other laboratories, that the impact of their discovery began to be appreciated by the scientific and technical community. What followed was an unprecedented worldwide outburst of excitement and activity and Bednorz and Müller were awarded the 1987 Nobel Prize in physics, making the time between a discovery and a Nobel Prize one of the shortest on record.

Since 1986, superconductivity in other cuprates has been discovered. The confirmed "record high" is $T_c \approx 125$ K (as of 1/91), well above the boiling temperature of liquid nitrogen (77 K). Superconductivity at such remarkably high temperatures *stirred the technological world with thoughts of commercial applications* and has attracted the attention of the public. Visions have been entertained of ore prospecting and brain-wave measurements with superconducting SQUIDs, long-distance lossless transmission lines, levitated trains, and lossless computer elements. Though some of the early euphoria has decreased, some of these applications may eventually become a reality. However, first, there is much to learn about these complex materials.

Such high values of T_c immediately cast doubts on the applicability of the conventional theories to the new materials. In the Bardeen-Cooper-Schrieffer (BCS) theory (or the strong-coupled BCS version) of **conventional superconductors** (defined as superconductors that do not contain Cu-O planes), electrons are bound in Cooper pairs by an electron-phonon interaction (i.e., phonon-mediated pairing). With only the BCS theory and the available phonon energies (of the order of the Debye energy), one was at first hard-pressed to account for such high T_c values. Since phonons are bosons, BCS-like theories with other kinds of bosons (e.g., electronic or spin excitations) as the "glue" were proposed.

There are numerous observations that support a BCS-like theory in some form or another. We list a few of these here.

1. In the superconducting state, the electrons are paired. Several different experiments (in 1987) showed that in the superconducting state, the fundamental charge is 2e, indicating paired electrons (Section 5-2a).

2. Photoemission spectroscopy, electron tunneling, and other experiments indicate an energy gap in the superconducting state (Chapter 5) just as in conventional superconductors. However, the energy gap is probably anisotropic and lies in the range $3.5k_BT_c$ to $8k_BT_c$, which is larger than the isotropic BCS value of $3.54k_BT_c$. However, with the strong-coupled BCS theory (Section 2-6), somewhat higher values can be understood.

3. Several experiments suggest pairing into a singlet-spin state and an s-wave orbital angular momentum state, as predicted by BCS. These experiments include single-particle and Josephson tunneling between a conventional superconductor (e.g., Pb or Sn) and some of the cuprate

superconductors (Section 5-2b), and also measurements of magnetic penetration depths.

4. These new high-T_c materials display many of the familiar superconducting properties, such as Josephson tunneling (Chapter 7) and vortex
structure, found in any type II superconductor (Chapter 6).

While the preceding aspects of the high-T_c cuprates are almost identical to conventional BCS-like properties, it is difficult to reconcile these
observations with other unusual properties. We list a few of these:

1. extremely high values of T_c

2. linear dc resistivity in the normal state

3. unusual behavior of the nuclear relaxation rate below T_c

4. close proximity of antiferromagnetic phases

5. extremely small coherence lengths

This combination of BCS and non-BCS-like properties brings forth
fundamental questions as to the nature of the normal state as well as that
of the superconducting state.

In addition to the high-T_c values, the large spatial anisotropy of these
materials is striking. The anisotropy is due to the layered crystal structure,
which, in the current thinking, is essential for the high-temperature
superconductivity. The layers are composed of Cu-O planes (or sheets),
separated from each other by planes of various other oxides and rare earths.
It is believed, on both experimental and theoretical grounds, that
superconductivity and charge transport are mostly confined to the Cu-O
planes.

A single Cu-O plane is sketched in Fig. 1-1a. Each Cu atom is surrounded by four O atoms in a square-planar configuration. In the high-T_c
structures, these Cu-O planes are **ab** planes perpendicular to the **c** axis. The
structures of $(La_{2-x}Sr_x)CuO_4$ and the other high-T_c crystals are discussed
at length in Chapter 3. Here, as an introduction, we only mention essential
structural details. In $(La_{2-x}Sr_x)CuO_4$, each Cu-O plane is separated from
its nearest-neighbor Cu-O planes (all perpendicular to the **c** axis) by ≈ 6.6
Å (Fig 1-1b), a fairly large distance in crystals. In between the two nearest
Cu-O planes are two La-O planes, indicated by dashed planes in Fig. 1-1b.
We use the shorthand notation La(n=1) to mean $(La_{2-x}Sr_x)CuO_4$, the
n=1 referring to one Cu-O plane, fairly isolated from the others. This notation is easily extendable to other high-T_c superconductors (Table 3-1) for
which Cu-O planes cluster in groups of two or three as indicated in Figs.
1-1c and 1-1d, respectively.

Doping is important for all of the high-T_c materials, since many can
be thought of as insulators that become metals only on the addition of ex-

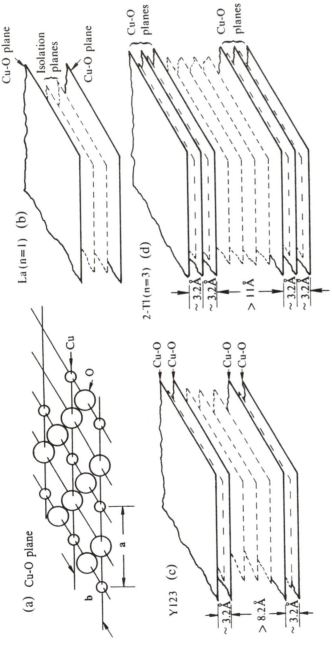

Fig 1-1 (a) An infinite plane of Cu-O atoms emphasizing the square-planar Cu-O₂ bonding. (b) A schematic diagram of the La(n=1) structure emphasizing the Cu-O planes (solid) perpendicular to the **c** axis, with two isolation La-O planes (dashed) between the Cu-O planes. (c) A schematic representation of the Y123 structure emphasizing the two immediately adjacent Cu-O planes. The sparsely occupied Y-atom plane between the two immediately adjacent Cu-O planes is indicated by a very light set of dashes. The three other (isolation) metal-O planes separating the pairs of Cu-O planes are indicated by heavier dashes. (d) A schematic representation of the structure of 2-Tl(n=3). Two sparsely occupied planes of Ca atoms between the three immediately adjacent Cu-O planes are indicated by very light dashes. The four other (isolation) metal-oxygen planes separating the three Cu-O planes are indicated by heavier dashes.

cess charge. For x=0, La(n=1) is an insulator. Doping with Ba, Sr, or Ca supplies only two electrons rather than the three supplied by La^{3+}; thus, these atoms act as acceptors. For $x \gtrsim 0.06$, the acceptor doping is large enough to make La(n=1) a metal and a superconductor with a maximum $T_c \approx 40$ K when $x \approx 0.15$. Doping in $YBa_2Cu_3O_{7-\delta}$ (Y123 in the accepted jargon) and many of the other high-T_c materials can be more subtle because they may be doped by a deficiency or excess of oxygen, or other atoms in the formula unit. This type of doping depends on the material processing, making the amount of excess charge less certain so that materials from different laboratories can differ slightly.

In early 1987, the discovery of $YBa_2Cu_3O_{7-\delta}$, with $T_c \approx 92$ K, was proof that the La material was not unique. Like La(n=1), Y123 has Cu-O planes perpendicular to the c axis, but it has two immediately adjacent Cu-O planes (i.e., n=2), about $3.2 \overset{\circ}{A}$ apart, separated from each other by a single, relatively sparsely occupied Y-atom plane (widely spaced dashes in Fig. 1-1c). In Y123, a pair of two immediately adjacent Cu-O planes is $>8.2 \overset{\circ}{A}$ distant from the next pair (Fig. 1-1c). Between one pair of immediately adjacent Cu-O planes and the next pair, three relatively unimportant metal-O planes are found, as indicated in Fig. 1-1c by dashed planes. These "unimportant planes" are sometimes referred to as "isolation planes" or "charge reservoirs." For $\delta \approx 1$, $YBa_2Cu_3O_{7-\delta}$ is an insulator, just like $(La_{2-x}Sr_x)CuO_4$ for x=0. However, for $\delta \lesssim 0.65$, Y123 is a metal and a superconductor.

The chemical formulae, T_c values, and shorthand notations for some of these high-T_c materials, with n values between 1 and 4, are listed in Table 3-1. The distance between the n immediately adjacent Cu-O planes is always small ($\approx 3.2 \overset{\circ}{A}$) and between these n Cu-O planes is a sparsely occupied Y or Ca plane. The sparsely occupied plane contains Y atoms in Y123, whereas Ca or Sr atoms are found in the Tl- and Bi-superconductor crystals. These n immediately adjacent Cu-O planes are always separated from the next n Cu-O planes by much larger distances. At present, the crystal $Tl_2Sr_2Ca_2Cu_3O_{10}$, 2-Tl(n=3), has the highest confirmed transition temperature, $T_c \approx 125$ K; it has sets of three immediately adjacent Cu-O planes, and separating one set from the next set are four metal-O isolation planes (Fig. 1-1d).

The structural anisotropy translates into an anisotropy of most physical properties, raising an important question. Do the observed properties have three-dimensional or perhaps two-dimensional character? In fact, the effective dimensionality may depend on temperature, because at low temperatures the in-plane superconducting coherence length is short ($\sim 3 \overset{\circ}{A}$), but near T_c it becomes extremely large. Characteristic length scales are discussed in Chapters 2 and 5. Of course, questions arise as to the effect of the anisotropy on the electron pairing mechanism.

In conclusion, it is clear that our current understanding of the high-T_c superconductors is incomplete, and fundamental questions remain. What is the pairing mechanism in high-T_c crystals and is it the same for all high-T_c materials? Is the pairing due to phonons with the addition of some other "booster" mechanism? If it is not phonon-mediated, do charge or magnetic excitations mediate the electron pairing? How can one explain the unusual normal-state properties? Do the charge carriers form a Fermi liquid in the normal state? Recent photoemission experiments indicate that there is a Fermi surface, but questions remain, since the correlations among charge carriers appear to be strong.

The field of high-temperature superconductivity is young and evolving rapidly. Although driven by experimental work, the theoretical effort has been immense. It is not yet clear whether the BCS theory can be sufficiently modified to survive as our basic understanding of this new superconductivity. It may be that radically new ideas, such as separation of spin and charge or the time/parity symmetry breaking of the fractional-statistics theory will be necessary to describe the normal and superconducting states. As a consequence of the intensely competitive scientific and technological efforts mounted in the high-T_c field, there are disagreements both in data and in theory. Some of these conflicts have been resolved and many others are being presently scrutinized by the researchers in the field. Bear in mind that the chemistry of these materials is complex. Thus, different processing steps can lead to materials with slightly different stoichiometries (and doping) and hence different properties. This book should be taken only as a progress report (to 1/91) in which we have attempted to focus on some basic aspects of this rapidly developing field. Reviews and papers that are more current should also be consulted.

The reader is assumed to have knowledge of the basic aspects of superconductivity covered in an introductory course in solid-state physics. In the next chapter, we review, and slightly expand on, a few of the topics particularly needed in the subsequent chapters. Essentially all of the effects studied in conventional superconductors are also found in the high-T_c materials. However, in the latter, some of these effects are exaggerated because of the higher temperatures, the small coherence lengths, and/or the anisotropy due to the Cu-O planar structure.

Problems

1. (a) Read the original Bednorz and Müller paper. Are you convinced of superconductivity? (See J.G. Bednorz and K. A. Müller, Z. Phys. B **64**, 189 (1986)). What simple, additional experiment might you have suggested? Hint: See Section 2-1.   (b) Read the original paper that reported superconductivity in Y123. Are you convinced of superconductivity? (See M. K. Wu, J. R. Ashburn, C. J. Torng, P. H. Hor, R. L. Meng, L. Gao, Z. J. Huang, Y. Q. Wang, and C. W. Chu, Phys. Rev. Lett. **58**, 908 (1987)).

2. (a) Read the original paper that reported laser action in ruby. Are you convinced of laser action? (See T. H. Maiman, Nature **187**, 493 (1960)). (b) Do the same for the original Mössbauer paper. Are you convinced of the importance of the Mössbauer effect? His Nobel Prize paper is in Science **137**, 731 (1962), which contains references to the initial work. (c) What conclusions might you draw from reading these papers?

Table 2-1 Values of T_c for the elements that are superconducting at atmospheric pressure. Many elements are superconducting under high pressure. Also listed are a very few of the known compounds, mostly emphasizing those with higher T_c values. $(SN)_x$ is a polymer. The last three compounds listed are heavy-electron metals. See the Notes.

Element	$T_c(K)$	Element	$T_c(K)$	Compound	$T_c(K)$
Nb	9.25	Al	1.17	Nb_3Ge	23.2
Tc	7.80	Ga	1.08	Nb_3Ga	20.3
Pb	7.20	Mo	0.915	Nb_3Al	18.6
$La\alpha$(hcp)	4.88	Am	0.85	Nb_3Sn	18.0
$La\beta$(fcc)	6.00	Os	0.66	Nb_3Au	10.8
V	5.40	Zr	0.61	V_3Si	17.1
Ta	4.47	Cd	0.517	V_3Ga	16.5
$Hg(\alpha)$	4.15	Ru	0.49	NbN	17.3
$Hg(\beta)$	3.95	Ti	0.40	MoC	14.3
Sn	3.72	Hf	0.128	Nb-Ti	10.3
In	3.41	Ir	0.113	$NbSe_2$	7.2
Tl	2.38	Lu	0.1	$(SN)_x$	0.26
Re	1.70	Be	0.026	UBe_{13}	0.85
Pa	1.40	W	0.0154	$CeCu_2Si_2$	0.65
Th	1.38	Rh	325×10^{-6}	UPt_3	0.54

Chapter 2

Review of Conventional Superconductors

So look to the starting point of the inquiry. See
whether it is satisfactorily stated, and try to answer
what I ask you as you think proper.

Plato, "The Crito"

As mentioned, the reader is assumed to have studied superconductivity at the level of most elementary solid-state physics (SSP) books. A few texts are listed in the Bibliography. In this chapter, some topics used in subsequent chapters will be reviewed and extended.

After an introduction, Sections 2-2 to 2-5 are devoted to phenomenological theories of superconductivity that give an understanding of the important length scales. Then, more microscopic theory is discussed.

2-1 Introduction

Heike Kammerlingh Onnes liquefied helium in 1908 (in Leiden), starting the field of low-temperature physics. Three years later, he reported another remarkable discovery. At 4.19 K, the resistance of mercury (Hg) dropped abruptly to zero. Thus, $T_c = 4.19$ K for Hg, and he found similar transitions in lead and tin to this new "superconducting" state.

Superconductivity is fairly commonplace among nonmagnetic metals. Of the elements, Nb has the highest transition temperature ($T_c = 9.25$ K). Table 2-1 lists T_c values for the elements and a few of the hundreds of compounds that have been reported.

In his Nobel Prize lecture (1913), Kammerlingh Onnes envisioned the use of superconductors for high-field electromagnets, because the power dissipation should be negligible. However, this idea had to wait until the

1960s, when type II superconductors were developed, before it could be implemented.

The most sensitive way of measuring a small resistance, to put an upper limit on the resistivity, is by means of a superconducting ring in which a **persistent current**, or a **supercurrent**, has been induced by magnetic induction. The supercurrent decay time can be measured by nuclear magnetic resonance techniques, for example. Only lower limits to the decay time have been found; 10^5 years has been given, but certain theoretical estimates yield times $\gg 10^{10}$ years. Upper limits of 10^{-23} Ω-cm have been experimentally set for the resistivity of a superconductor, which can be compared to 10^{-9} Ω-cm for pure copper.

In 1933, it was found that when a superconductor was cooled below T_c in a magnetic field, the magnetic flux was expelled. Thus, in a weak magnetic field, a superconductor has **perfect diamagnetism**, a phenomenon called the **Meissner effect**. Perfect diamagnetism is completely different from what one would expect in a metal with infinite conductivity, which, when cooled in a magnetic field, would trap the magnetic flux in the conductor. The Meissner effect implies that in a magnetic field, superconductors develop surface current, which give rise to magnetic fields that exactly cancel the external field, leaving a field-free bulk. More will be said about the penetration depth of these surface currents. The Meissner effect also implies a **critical field**, H_c, above which superconductivity will be destroyed. Since $H^2/8\pi$ is the magnetic energy density, the difference between the Helmholtz free energy in the normal and superconducting states per unit volume is

$$F_n(T) - F_s(T) = H_c^2/8\pi \qquad (2-1a)$$

where T is temperature. Experimental values of H_c vs. T are shown in Fig. 2-1. This phase diagram indicates that the material is superconducting for field and temperature values below and to the left of any curve, while above and to the right of a curve, the material is in its normal state. The experimental results can be approximately described by a quadratic temperature dependence,

$$H_c(T) = H_c(0) \left[1 - (T/T_c)^2\right] \qquad (2-1b)$$

with $H_c(0)$ values less than 10^3 Oe for the elements. A few representative $H_c(0)$ values are given in Table 2-2, which are obtained from data of the type in Fig. 2-1.

Thus, a superconductor can be characterized by perfect conductivity and perfect diamagnetism. This is different from a "perfect conductor," which

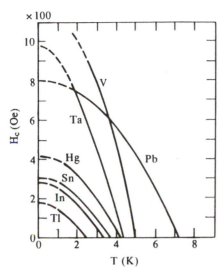

Fig. 2-1 H_c vs. T for a few elemental superconductors. The sample is normal for $H > H_c(T)$ and superconducting for $H < H_c(T)$.

would not have the latter property. Actually, in normal metals, Bloch waves theoretically have perfect conductivity (but not perfect diamagnetism), and resistance occurs only because the electrons are scattered by deviations from the perfect periodicity of the crystal structure. Deviations may be due to phonons, impurities, or other imperfections (Section 4-3).

Even in type I superconductors, there are geometrical (sample shape-dependent) effects that may cause the magnetic field to exceed H_c in some parts of the sample, while $H < H_c$ in other parts of the sample. This is the **intermediate state**, which is not to be confused with the **mixed state** or **vortex state** found in type II superconductors. In the intermediate state, the fraction of the sample in the normal state naturally increases as H increases and is of macroscopic size. In type II superconductors, the dimension of the vortex is microscopic and determined by the coherence length.

2-2 Two-Fluid Model

The superfluid properties of helium (^{4}He, which has zero electron and nuclear spin and thus is a boson) can be well understood by the two-fluid model. The helium atoms can be considered to be in two states. A fraction of the atoms are in the condensed Bose-Einstein ground state, while the rest are in the "normal" state. The fraction in the condensed state are assumed to lead to the remarkable properties of superfluid He.

Table 2-2 Properties of selected superconductors. H_c at 0 K is given for the elemental materials and H_{c2} at 0 K for the compounds, which are all type II superconductors. While the units of H_c are mT, those of H_{c2} are 10^4 Oe = 1 T. $N(E_F)$ is the density of states at the Fermi energy and its units are states/(atom-eV); it is obtained from the linear term in the normal-state specific heat (Eq. 2-6f). See Kinoshita (Bib.) for more extensive lists of data.

Mat.	T_c (K)	Θ_D (K)	H_c (mT)	$N(E_F)$	$\dfrac{2\Delta(0)}{k_B T_c}$	λ_{ep}
Nb	9.25	276	206	0.91	3.6	0.85
Tc	7.80	411	141	0.62	3.6	0.65
Pb	7.20	96	80.3	0.28	4.5	1.55
V	5.40	383	141	1.31	3.4	1.0
Ta	4.47	258	83	0.77	3.5	0.75
Hg	4.19	87	41.1	0.15	4.6	1.6
			H_{c2}			
NbN	17.3	309	47	0.45	4.3	1.06
ZrN	10.7	349	0.3	0.34	-	0.67
MoC	14.3	620	9.8	0.53	-	0.67
NbC	11.6	355	2.0	0.33	-	0.72
Nb_3Ge	23.2	302	38	0.95	4.2	1.80
Nb_3Ga	20.3	-	34	1.18	4.1	1.74
V_3Ga	16.5	310	27	2.7	4.0	1.17
V_3Si	17.1	330	25	2.5	3.6	1.10
HfV_2	9.4	187	20	2.05	-	1.00
$PbMo_6S_8$	15.3	411	60	0.67	3.84	1.20
$PbMo_6Se_8$	6.7	294	7	0.24	-	0.66

Gorter and Casimer (1934) used this idea of superfluid helium and applied it to superconductivity. The conduction electron density is $n = N/V$, where N is the number of conduction electrons in the sample of volume V. Then n_n and n_s are the densities of normal-state and superconducting electrons, where $n = n_n + n_s$. Of course, the separation of the conduction electrons in this manner is a drastic assumption. Take x ($= n_n/n$) and $1-x$ to be the fractions of normal-state and superconducting-state electrons, respectively. They assumed a free energy for the conduction electrons of the form,

$$F(x, T) = x^{1/2} f_n(T) + (1 - x) f_s(T) \qquad (2 - 2a)$$

The f_n and f_s terms were taken as

$$f_n(T) = - \gamma T^2/2$$
$$f_s(T) = - \beta \quad \text{(a constant)} \tag{2 - 2b}$$

The $\gamma T^2/2$ term is the usual free-electron energy in a normal metal that yields a γT (linear) specific heat at low temperatures. (See the Problems.) The superconducting condensation energy is taken as $- \beta$. At $T=0$, the free energy is $- \beta$, since all the electrons are in the condensed state, and at $T = T_c$, it is $- \gamma T_c^2/2$, since $x=1$.

Minimization of $F(x, T)$ with respect to x, for fixed T, yields

$$x = n_n/n = (T/T_c)^4 \tag{2 - 2c}$$

Thus, there is a sharp temperature dependence of the fraction of normal state electrons just below T_c.

Using the thermodynamic relation for the difference between the free energies in the normal and superconducting states in terms of $H_c^2/8\pi$ (Eq. 2-1a) and using Eqs. 2-2a to 2-2c, then the quadratic expression for $H_c(T)$ in Eq. 2-1b is obtained. Additionally, relating the entropy to the temperature derivative of free energy and the specific heat to the temperature derivative of the entropy, the electronic specific heat is

$$C_{es} = 3\gamma T_c(T/T_c)^3 \tag{2 - 2d}$$

Thus, at T_c, the ratio of the electronic specific heat in the superconducting and normal phases is $C_{es}/C_{en} = 3$, which is in reasonable agreement with experiment.

These agreements with experiment are not too surprising, since the unusual form for the free energy (Eq. 2-2a) was chosen to yield these results. Nevertheless, the two-fluid model gives a physical basis for understanding superconductivity, a useful free energy expression that yields quantities in agreement with experiment, and $x = n_n/n = (T/T_c)^4$, which will be used later.

2-3 London Equation

The brothers F. and H. London (1935) used ideas based on the two-fluid (Section 2-2) to try to understand the Meissner effect. Let n, n_n, n_s be the densities of all the conduction electrons, the normal state electrons, and the superconducting electrons, with $n = n_n + n_s$. Assume $n_s(T_c) = 0$

and $n_s(0) = n$. Then, the current due to the superconducting electrons is given by $J = - ev_s n_s$ and from Newton's law, $m(dv/dt) = -eE$,

$$\frac{\partial J}{\partial t} = (\frac{e^2 n_s}{m}) E$$

$$E = \frac{\partial}{\partial t} (\Lambda J) \qquad (2 - 3a)$$

where $\qquad \Lambda \equiv m/n_s e^2$

Combining these results with Maxwell's equation, $\nabla \times E = - c^{-1}(\partial B/\partial t)$, yields

$$\frac{\partial}{\partial t} [c\nabla \times (\Lambda J) + B] = 0 \qquad (2 - 3b)$$

This is a general equation for any metal with conduction electron density n_s and will not account for the Meissner effect. The Londons realized that the characteristic behavior of a superconductor could be obtained by restricting the full set of solutions of Eq. 2-3b to those where the expression within the square bracket in Eq. 2-3b is zero, not only its time dependence. Thus,

$$B = - c\nabla \times (\Lambda J) \qquad (2 - 3c)$$

which is the **London equation**, and Λ or n_s can be considered a phenomenological parameter. Taking the curl of both sides, using Maxwell's equation, $\nabla \times B = (4\pi/c)J$, and using the identity, $\nabla \times \nabla \times B = \nabla(\nabla \cdot B) - \nabla^2 B = - \nabla^2 B$, we obtain

$$\nabla^2 B = B/\lambda_L^2 \qquad \nabla^2 J = J/\lambda_L^2$$

$$\lambda_L^2 \equiv mc^2/4\pi n_s e^2 \qquad (2 - 3d)$$

where the **London penetration depth** λ_L is defined. Equations 2-3d indicate that at an air-superconductor interface, the magnetic field decays from the surface into a superconductor bulk exponentially. Letting the superconductor surface be at right angles to the x axis with positive x in the superconductor, the solutions are

$$B_z = B_z(0) \exp(- x/\lambda_L)$$

$$J_y = cB_z(0)/4\pi\lambda \exp(- x/\lambda_L) \qquad (2 - 3e)$$

where $B_z(0)$ is the magnetic field at the interface. Thus, the London equation gives a simple picture of the Meissner effect; a current is set up that shields the interior of the sample from the external magnetic field.

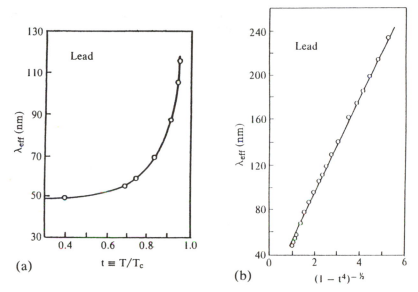

Fig. 2-2 (a) λ vs. reduced temperature $t \equiv T/T_c$. For details, see F. F. Mende, A. L. Spitsyn, N. A. Tereschenko, and O. E. Rudnev, Sov. Phys. Tech. Phys. **22**, 1111 (1977). (b) The temperature dependence of λ in lead plotted for ease of comparison to theory. For details, see R. F. Gasparovic and W. L. McLean, Phys. Rev. B **2**, 2519 (1970).

No intrinsic temperature is predicted other than that from $n_s(T)$. We can use the two-fluid result for the temperature dependence of n_s (Eq. 2-2c)

$$1 - x = n_s(T)/n = 1 - (T/T_4)^4 \qquad (2-3f)$$

Combining this result with the expression for λ_L^2 (Eq. 2-3d), we find

$$\lambda_L(T) = \frac{\lambda_L(0)}{[1 - (T/T_4)^4]^{1/2}} \qquad (2-3g)$$

Thus, no flux is excluded at T_c, since $\lambda = \infty$. However, below T_c, λ_L decreases rapidly, leading to the exclusion of flux from the bulk of the sample, the Meissner effect. Experimental $\lambda(T)$ values in lead are shown in Fig. 2-2a; note the rapid decrease below T_c. Reasonable agreement of $\lambda(T)$ with the form given in Eq. 2-3f can be seen in Fig. 2-2b, although deviations are found in other materials.

The London equations can be reformulated in a more intuitive manner, in terms of the vector potential **A**, which is useful when discussing Ginzburg-Landau theory. The canonical momentum is $\mathbf{p} = m\mathbf{v} + e\mathbf{A}/c$. In zero field, the ground state should have zero net momentum. In the presence of a magnetic field, the average velocity of the superconducting particles should be proportional to the first power of the field,

$$< v_s > = - eA/mc \qquad (2 - 3h)$$

Since n_s is the number density of superconducting electrons, then

$$J_s = n_s e < v_s > = - n_s e^2 A/mc$$
$$= A/\Lambda c \qquad (2 - 3i)$$

where Λ is defined in Eq. 2-3a.

Equation 2-3i is not gauge-invariant, so can only be correct for a particular choice of gauge. The choice is the **London gauge** specified by div $A=0$ (so that div $J = 0$), the normal component of A over a surface is related to the supercurrent through the surface by Eq. 2-3i, and A must approach zero in the interior of the superconductor.

Given the definition, the penetration depth, at 0 K, it should be expected that $n_s = n$, the total number of conducting electrons. Thus, using Eq. 2-3d,

$$\lambda_L(0) = (mc^2/4\pi ne^2)^{\frac{1}{2}} \qquad (2 - 3j)$$

corresponding to values 100-1000Å. Measurements of the penetration depth, in elemental superconductors, tend to give penetration depths larger than this calculated $\lambda_L(0)$ value. This discrepancy can be understood as arising from nonlocal electrodynamics.

2-4 Nonlocal Fields

2-4a Nonlocal Electrodynamics Sketched — Pippard (1953) measured the penetration depth in alloys and found that it was more sensitive to the amount of alloying than the thermodynamic superconducting properties, H_c and T_c. He deduced that the penetration depth is a sensitive function of the electron **mean free path**, ℓ, which varies rapidly with alloying. For example, tin with 3% indium had a penetration depth a factor 2 larger than pure tin, yet H_c and T_c changed by only a few percent.

To explain these data, Pippard proposed the use of nonlocal electrodynamics where, for example, the current at a point depends on the electric field not at just the same point but throughout a volume. Pippard used Chambers nonlocal relation, in which the current and electric field at the origin, $J(0) = \sigma E(0)$, are replaced by

$$J(0) = \frac{3\sigma}{4\pi\ell} \int \frac{r(r \cdot E) \exp(-r/\ell)}{r^4} \, d^3r \qquad (2-4a)$$

so that the current depends on the field throughout a volume of radius $\sim\ell$ about the origin. Pippard argued that the superconducting wave function also should have a characteristic length, ξ_0, which would play a role analogous to an electron mean free path in a normal metal. Further, ξ_0 could be estimated via the Uncertainty Principle. He argued that only electrons within $\sim k_B T_c$ of E_F (the Fermi energy) should be important in superconductivity. Since $p^2/2m$ is the energy of a free electron, the important electrons have a momentum range $\Delta p \sim k_B T_c / v_F$, where v_F is the Fermi velocity. From the Uncertainty Principle, $\Delta x \sim \hbar/\Delta p \sim \hbar v_F / k_B T_c$. Thus, a characteristic length can be defined as

$$\xi_0 = \alpha \frac{\hbar v_F}{k_B T_c} \qquad (2-4b)$$

where α is of the order of unity. BCS obtain d $\alpha = 0.18$.

Borrowing an idea from the mean free path in a normal-metal alloy, in the presence of impurity scattering, the coherence length ξ might be expected to be related to ℓ by an equation of the type,

$$\frac{1}{\xi} = \frac{1}{\xi_0} + \frac{1}{\ell} \qquad (2-4c)$$

where ξ_0 is the coherence length of the pure material. Normally, $\xi_0 > \ell$ which is the **clean limit**. Increased alloying causes increased electron-impurity scattering, leading to $\ell \ll \xi_0$, then $\xi \approx \ell$, the **dirty limit**. Using experimental data from both tin and aluminum, and computing the penetration depth for various values of ξ_0 and λ, it was found that $\alpha = 0.15$ (Eq. 2-4b) fit the data. Further, a penetration depth considerably larger than λ_L was found.

Thus, Pippard's nonlocal electrodynamics brings out two length scales, a coherence length ξ, as well as a magnetic penetration depth λ.

2-4b Various Situations and Dirty Superconductors — Most conventional superconductors can be classified in the following ways.

(a) A pure superconductor with large intrinsic coherence length ξ_0 (large v_F or small T_c, Eq. 2-4b). In this case, Pippard has shown that

$$\lambda = 0.65\lambda_L(\xi_0/\lambda_L)^{1/3} \quad \text{for } \xi^3 \gg \xi_0\lambda_L^2 \qquad (2-4d)$$

Since ξ_0/λ can be quite large, λ can be considerably larger than λ_L.

(b) An impure or alloyed superconductor where the electron mean free path, ℓ, is much smaller than ξ_0 (a **dirty superconductor**). This is called the **dirty limit**, $\xi \approx \ell$ and $\xi \ll \xi_0$. (See Section 2-5f.) In this case,

$$\lambda = \lambda_L (\xi_0/\xi)^{\frac{1}{2}} \approx \lambda_L (\xi_0/\ell)^{\frac{1}{2}} \quad \text{for } \xi^3 \ll \xi_0 \lambda_L^2 \qquad (2-4e)$$

Thus, λ can be considerably larger than λ_L.

(c) Pure metals with small ξ_0. In this case, $\xi_0 \ll \lambda_L$ and the penetration depth is λ_L. However, if ℓ is also very small, Eq. 2-4e should be used.

2-5 Ginzburg-Landau Theory

2-5a   Introduction — Ginzburg and Landau proposed a phenomenological theory of superconductivity (1950), which is related to Landau's theory of second-order structural phase transitions. The free energy is expanded in terms of an order parameter, which is zero in the high-temperature phase and describes the "ordered" electrons in the low-temperature phase. The Ginzburg-Landau, or **GL theory**, introduces a complex pseudo-wave function ψ as the order parameter, which is hypothesized to be related to the *local density of superconducting electrons* as

$$n_s \equiv N_s/V = |\psi(\mathbf{r})|^2 \qquad (2-5a)$$

and n $(\equiv N/V)$ is the conduction band *electron density*.

Using the GL formalism, they were able to treat nonlinear (in magnetic fields) and spatial variation effects of n_s. Besides a temperature-dependent penetration depth (λ), they also obtain a temperature-dependent coherence length (ξ). The latter characterizes the distance over which $\psi(\mathbf{r})$ can vary without a substantial energy increase. The important **GL parameter** is the ratio of these two lengths,

$$\kappa \equiv \lambda/\xi \qquad (2-5b)$$

Elemental superconductors have values, $\lambda \sim 500\text{Å}$ and $\xi \sim 4000\text{Å}$, so $\kappa \ll 1$ for these materials. Figures 2-3a and 2-3b show the normal-superconducting boundary in the intermediate state (Section 2-1), how the magnetic field penetrates the superconductor (to a depth λ), and how ψ increases in the superconductor to its value at infinity ψ_∞ (in a distance ξ). However, the importance of the $\kappa \gg 1$ case was not fully appreciated until Abrikosov (1957) showed how this leads to type II superconductors. Essentially all superconducting compounds and all high-T_c materials have $\kappa \gg 1$.

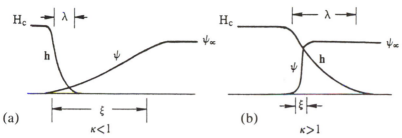

Fig. 2-3 Schematic diagrams showing how the local magnetic field and $\psi(x)$ vary with the distance from a normal-superconductor interface. The two cases, (a) and (b), show the GL parameter much smaller and larger than unity, as indicated.

When first proposed, the phenomenological GL theory was not thought to be particularly important. However, it turns out that it is an extraordinary example of intuitive physics. One of the very important BCS (1957) results is the occurrence of a superconducting gap, $2\Delta(T)$, in the allowed electron states below T_c. Gor'kov (1959) showed that *the GL theory was a limiting form near T_c of the BCS theory* when BCS is generalized to deal with spatially varying situations. Gor'kov showed that $\psi \propto \Delta(\mathbf{r})$, which for a homogeneous medium reduces to the BCS gap parameter, and that ψ can be thought of as the wave function of the center of mass of a Cooper pair. Today, the GL theory is not only appreciated, it is essentially the only way to deal with spatially inhomogeneous systems such as thin films, proximity systems (Notes), and others. The GL theory is extensively used for analyzing superconductors in a magnetic field because of the spatial inhomogeneities that occur due to penetration and the intermediate state in type I superconductors, and the mixed state in type II superconductors (Section 2-5f).

The Ginzburg-Landau theory is complicated and has been applied to many situations. We outline the "bare bones" of the GL theory in the phenomenological way as originally proposed by them. The more interested reader should consult the Bibliography.

2-5b GL Free Energy — First, consider the GL theory for the simple case in the absence of magnetic fields and in the absence of spatial variations. Then the free energy density between the superconducting and normal state is taken as

$$f_s - f_n = \alpha |\psi|^2 + \tfrac{1}{2}\beta |\psi|^4 \qquad (2-5c)$$

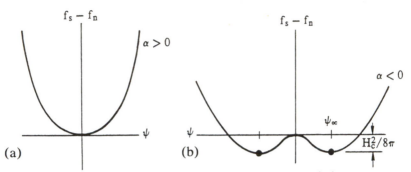

Fig. 2-4 The GL free energy for no gradients or fields (Eq. 2-5c) vs. $|\psi|$ for (a) positive and (b) negative α. The equilibrium values of $|\psi|$ are 0 and $|\psi_\infty|$, respectively.

where α and β are parameters to be determined. Using $n_s = |\psi|^2$ from Eq. 2-5a, the equilibrium value of the order parameter is obtained from $\partial(\Delta f)/\partial n_s = 0$. For $\alpha > 0$, the minimum of the free energy is at $|\psi|^2 = 0$, the normal state. However, for $\alpha < 0$,

$$|\psi|^2 = -\alpha/\beta \equiv |\psi_\infty|^2 \qquad (2-5d)$$

then
$$f_s - f_n = -H_c^2/8\pi = -\alpha^2/2\beta$$

The free energy vs. ψ is shown in Fig. 2-4 for α positive $(T > T_c)$ and negative $(T < T_c)$. In Eq. 2-5d, ψ_∞ is defined as the ψ value in the interior of the sample, far from any gradients in this parameter. H_c is the thermodynamic critical field (Fig. 2-1). With the help of Fig. 2-4, the free energy form in Eq. 2-5c should be clearer. At T_c, there must be a change in the sign of α, and β must be positive for a second-order phase transition. Defining the **reduced temperature** as $t \equiv T/T_c$, expanding $\alpha(T)$ about T_c, and keeping only the leading term, we write

$$\alpha(T) = \alpha'(t-1) \qquad \alpha' > 0 \qquad (2-5e)$$

Relatedly, we take β to be free from singularities or zeros near T_c. These temperature dependencies for α and β are also consistent with $H_c(T) \propto 1 - t$ very near T_c from Eq. 2-5d. Note that just below T_c, these temperature dependencies also yield

$$|\psi|^2 \propto n_s \propto (1-t) \qquad (2-5e)$$

where the relations in Eq. 2-5a are repeated. These GL results only apply very close to T_c. Thus, while over the entire temperature range from zero to T_c we expect $H_c(T) \propto (1 - t^2)$ (Eq. 2-1b) and $n_s \propto (1 - t^4)$ (Eq. 2-2c), close to T_c we find that $H_c(T) \propto (1 - t)$ and $n_s \propto (1 - t)$ is a good approximation.

Now consider the full GL free energy in the presence of magnetic fields and spatial gradients. The free energy density is taken as

$$f = f_n + \alpha |\psi|^2 + \frac{\beta}{2} |\psi|^4$$
$$+ \frac{1}{2m^*} |(\frac{\hbar}{i} \nabla - \frac{e^*}{c} A)\psi|^2 + \frac{h^2}{8\pi} \qquad (2-5c)$$

where the free energy density in the normal state ($\psi=0$) is $f_n(T) = f_{n0}(0) - \frac{1}{2}\gamma T^2$; the second term leads to the linear-temperature specific heat for a metal (Section 2-6d) and $h^2/8\pi$ is the classic magnetic energy density. We follow the convention and let $\mathbf{h(r)}$ be the **local microscopic field**. Then $\mathbf{B}$, which is $=\mathbf{H}+4\pi\mathbf{M}$ or $=\mu_0(\mathbf{H} + \mathbf{M})$ in SI units, is the macroscopic average of $\mathbf{h(r)}$. For superconductors, $\mathbf{B} \approx \mathbf{H}$, since $\mathbf{M}$ is very small, typically much smaller than the applied field. However, $\mathbf{h(r)}$ may differ considerably from $\mathbf{B}$, varying considerably within the unit cell, vortex, and so forth.

This form of the free energy density (Eq. 2-5c) is taken for temperatures near T_c, where ψ is small. The extra term in ψ can be associated with the kinetic energy associated with extra wiggles in describing how n_s varies in space. If n_s varies slowly in space, it should be sufficient to keep only the leading term $|\text{grad}\psi|^2$, which is combined with the vector potential $\mathbf{A(r)}$ in a gauge-invariant manner. This expansion, in terms of even powers of the order parameter, is also common in studies of structural phase transitions where gradients of the order parameter lead to domain effects.

Now, further consider the terms in the GL free energy involving gradients and fields. Taking $\psi = |\psi| e^{i\phi}$ to give the order parameter a magnitude and phase, from this part of the free energy we obtain

$$\frac{1}{2m^*} \left[\hbar^2(\nabla |\psi|)^2 + \left(\hbar\nabla\phi - \frac{e^*}{c} A \right)^2 |\psi|^2 \right] \qquad (2-5f)$$

The first term involves gradients of the magnitude of the order parameter. For example, these would be nonzero at a domain wall. The second term is a kinetic-energy-like term, which is associated with the supercurrent. In the London Gauge, ϕ is constant, so this term is $(e^{*2}A^2/2m^*c^2)|\psi|^2$. Using the London approach with the vector potential $\mathbf{A}$, the energy density of a superconductor is $A^2/8\pi\lambda_{eff}^2$, where a general penetration depth is written as λ_{eff}. Equating this energy density to that found from the GL equations, we obtain a relation for λ_{eff}^2,

$$\lambda_{eff}^2 = m^*c^2/(4\pi e^{*2}|\psi|^2) \qquad (2-5g)$$

Using $n_s = |\psi|^2$, the usual London penetration depth (Eq. 2-3i) is obtained, except for the starred quantities, m^* and e^*.

Anticipating electron pairing, we can deal with the starred quantities by taking $e^*=2e$ and $m^*=2m$, then $n_s^*=\frac{1}{2}n_s$ where n_s is the ordinary (nonpaired) electron density. With these substitutions in Eq. 2-5g, λ_{eff} is unchanged with pairing.

Taking the above values for the starred quantities, Eqs. 2-5e and 2-5g can be combined to yield

$$|\psi_\infty|^2 \equiv n_s^* \equiv \frac{n_s}{2} = \frac{m^*c^2}{4\pi e^{*2}\lambda_{eff}^2} = \frac{mc^2}{8\pi e^2\lambda_{eff}^2}$$

$$\alpha(T) = -\frac{e^{*2}}{m^*c^2} H_c^2(T)\lambda_{eff}^2(T) = -\frac{2e^2}{mc^2} H_c^2(T)\lambda_{eff}^2(T) \qquad (2-5h)$$

$$\beta(T) = \frac{4\pi e^{*4}}{m^{*2}c^2} H_c^2(T)\lambda_{eff}^4(T) = \frac{16\pi e^4}{m^2c^4} H_c^2(T)\lambda_{eff}^4(T)$$

The temperature dependence of the quantities in Eq. 2-5h can be obtained, since they only involve $\lambda_{eff}(T)$ and $H_c(T)$. Using $t \equiv T/T_c$, the **reduced temperature**, and the experimental $H_c \propto (1-t^2)$ and $\lambda^{-2} \propto (1-t^4)$, from Eqs. 2-1b and 2-3g,

$$|\psi_\infty|^2 \propto 1 - t^4 \sim 1 - t$$

$$\alpha \propto -\frac{1-t^2}{1-t^4} \sim t - 1 \qquad (2-5i)$$

$$\beta \propto \frac{1-t^2}{(1-t^4)^2} \sim 1$$

As can be seen, α is positive for $T > T_c$ and negative for $T < T_c$, as expected (Fig. 2-4).

2-5c GL Differential Equations — In the presence of fields, currents, and gradients, $\psi(\mathbf{r}) = |\psi(\mathbf{r})| \exp i\phi(\mathbf{r})$ adjusts so as to minimize the total free energy. Thus, setting the variation of the free energy with respect to the order parameter equal to zero, we obtain

$$\alpha\psi + \beta|\psi|^2\psi + \frac{1}{2m^*}\left(\frac{\hbar}{i}\nabla - \frac{e^*}{c}\mathbf{A}\right)^2\psi = 0 \qquad (2-5j)$$

Along with this equation, GL wrote the usual quantum expression for the current as

$$\mathbf{J} = \frac{c}{4\pi}\nabla \times \mathbf{h} = \frac{e^*\hbar}{2m^*i}(\psi^*\nabla\psi - \psi\nabla\psi^*) - \frac{e^{*2}}{m^*c}\psi^*\psi\mathbf{A}$$
$$= \frac{e^*}{m^*}|\psi|^2(\hbar\nabla\phi - \frac{e^*}{c}\mathbf{A}) = e^*|\psi|^2\mathbf{v}_s \qquad (2-5j)$$

Equations 2-5j are referred to as the GL differential equations. The first has a form similar to that of the Schrödinger equation for a particle of charge e*, mass m*, and an energy eigenvalue $-\alpha$. With the use of the GL differential equations, a characteristic length can be defined (Section 2-5e).

2-5d Flux Quantization — A striking result can be obtained from the **J** expression in Eq. 2-5j. Consider a superconducting disk with a hole in it. Integrate **J** around the hole along a path deep in the superconductor, far from the surface shielding currents. Then

$$0 = \int \mathbf{J}\cdot d\mathbf{l} = \int(\hbar\nabla\phi - \frac{e^*}{c}\mathbf{A})\cdot d\mathbf{l} \qquad (2-5k)$$

where d**l** is the path of integration. Using Stoke's theorem,

$$\int \mathbf{A}\cdot d\mathbf{l} = \int \nabla \times \mathbf{A}\cdot d\mathbf{S} = \int \mathbf{B}\cdot d\mathbf{S} = \Phi \qquad (2-5l)$$

where Φ is the flux enclosed in the hole. Now consider integration of the phase. Since $\psi(\mathbf{r})$, the order parameter, must be single-valued (Bohr condition), its phase must change an integer number times (n) 2π in going around the hole. Thus,

$$\int \nabla\phi\cdot d\ell = \Delta\phi = 2\pi n \qquad (2-5l)$$

Combining these results, the quantization of flux is obtained,

$$\Phi = n\frac{hc}{e^*} = n\frac{hc}{2e} \equiv n\,\Phi_0 \qquad (2-5m)$$

where e*=2e has been inserted, but at the time of the original work, electron pairing was not appreciated. The quantity $\Phi_0 = hc/2e$ = 2.0679×10^{-7} gauss-cm^2 is called the **fluxoid** or **flux quantum**. Measurement of the flux quantum (1961) demonstrated that e*=2e, showing that *paired electrons* occur in the superconducting state. The measurement also shows that the coherence of this quantum mechanical state extends over macroscopic dimensions. Actually, London first obtained Eq. 2-5m for the fluxoid, naturally also without appreciating the significance of e*.

2-5e   GL Coherence Length — Consider the GL differential equations (Eq. 2-5j) in the absence of fields ($A = 0$). Since all of the coefficients are real, ψ may be taken to be real, and let $G \equiv \psi/\psi_\infty$, where $\psi_\infty^2 = -\alpha/\beta$ (Eq. 4-5d). Then, in one dimension, Eq. 2-5j becomes

$$\frac{\hbar^2}{2m^*|\alpha|}\frac{d^2G}{dx^2} + G - G^3 = 0$$

$$\xi^2(d^2G/dx^2) + G - G^3 = 0 \qquad (2-5n)$$

where, from dimensional arguments, a characteristic length $\xi(T)$ can be defined below T_c as

$$\xi^2(T) \equiv \hbar^2/2m^*|\alpha| \qquad (2-5o)$$

This length describes the variation of ψ due to a small disturbance, as is evident from investigating Eq. 2-5n. To do this, let $G = 1+g$, with $g \ll 1$. Then to first order in g,

$$\xi^2(d^2g/dx^2) + (1+g) - (1+3g+...) = 0$$

$$d^2g/dx^2 = (2/\xi^2)g \qquad (2-5p)$$

$$g(x) \sim \exp \pm [\sqrt{2}\, x/\xi(T)]$$

Thus, a small disturbance of ψ will decay to ψ_∞ with a characteristic length $\xi(T)$.

For the coherence length, $\xi(T)$ (Eq. 2-5o), we have used the same symbol as Pippard's ξ_0 (Eq. 2-4b), yet the relation between the two lengths is not apparent. Their relation can be found using Eq. 2-5h for α,

$$\xi(T) = \Phi_0/(8)^{1/2}\pi H_c(T)\lambda_{eff}(T) \qquad (2-5q)$$

where Φ_0 is the fluxoid. BCS finds that the fluxoid can be related to ξ_0, the 0 K London penetration depth, and thermodynamic critical field as $\Phi_0 = (2/3)^{1/2}\pi^2\xi_0\lambda_L(0)H_c(0)$. Thus, very near T_c, in the pure and dirty limits,

$$\xi(T)_{pure} = 0.74\xi_0/(1-t)^{1/2}$$

$$\xi(T)_{dirty} = 0.86(\xi_0\ell)^{1/2}/(1-t)^{1/2} \qquad (2-5r)$$

The important dimensionless **Ginzburg–Landau parameter** shown in Eq. 2-5b is

$$\kappa \equiv \frac{\lambda_{eff}(T)}{\xi(T)} = 8^{\frac{1}{2}}\pi H_c(T)\lambda_{eff}^2(T)/\Phi_0 \qquad (2-5s)$$

Using the experimental results for $H_c(T)$ and $\lambda_{eff}(T)$, as for Eq. 2-5i, $\kappa \sim 1 + t^2$ near T_c, showing that κ is not singular near T_c.

2-5f   Type II Superconductors — Essentially all pure, elemental superconductors have $\xi \gg \lambda_{eff}$, so $\kappa \ll 1$ (Fig. 2-3a). However, **alloying** or forming compounds causes the electron mean free path ℓ to be much smaller than ξ_0 (a **dirty superconductor**) hence (Eq. 2-4c) $\xi \approx \ell$. Thus, alloying a pure superconductor will turn it into a material with the opposite limit, $\kappa \gg 1$. These two κ values lead to positive and negative surface energies, respectively, between a normal and superconducting domain (e.g., in the intermediate state). The crossover from positive to negative surface energy occurs for $\kappa = 1/\sqrt{2}$, as shown in the original GL paper.

It was not until the work of Abrikosov (1957), however, that the full implications of the $\kappa > 1/\sqrt{2}$, negative surface-energy regime was appreciated. The negative surface energy causes the normal-state, flux-bearing regions to subdivide until a single fluxoid quantum passes through, like a tube, in the otherwise superconducting sample. The tube of flux is called a **filament** or **vortex**. This is **type II superconductivity**, also called the **mixed state** or **vortex state**. The behavior of type I superconductors ($\kappa < 1/\sqrt{2}$) below H_c, and type II superconductors for $H_{c1} < H < H_{c2}$, is discussed in most elementary SSP books; we only discuss a few special points.

The magnetic-phase diagram for a type II superconductor is shown in Fig. 2-5a (which may be compared to that for a type I superconductor in Fig. 2-1). Above the $H_{c2}(T)$ curve, the sample is normal. For $H_{c1} < H < H_{c2}$, the superconductor is in the vortex state. Below the $H_{c1}(T)$ curve, all the flux is expelled from the bulk as in a type I superconductor, leading to what is sometimes called the **Meissner state**.

The **vortex energy per unit length** is found to be

$$E_V/L = [(n\Phi_0)^2/4\pi\lambda^2] \ln \kappa \propto (n\Phi_0)^2 \qquad (2-5t)$$

Thus, a vortex containing n flux quanta would have a higher energy than n vortices, each containing one flux quantum. The latter is experimentally observed.

Relations between the critical fields, H_{c1} and H_{c2}, and length scales in type II superconductors can be estimated. When the applied field is just above H_{c1}, the field at the vortex core will be $\sim H_{c1}$, radially decreasing to

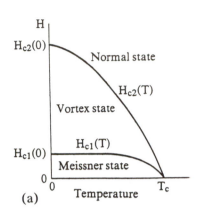

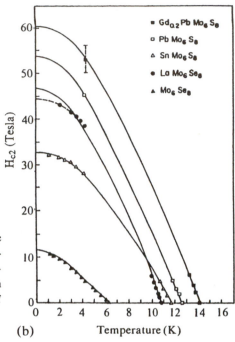

(b)   Temperature (K)

Fig. 2-5 (a) The mean field magnetic phase diagram for a type II superconductor. (b) H_{c2} vs. T for the materials indicated. H_{c1} is too small to be shown on this plot. For details see O. Fischer, J. of Physics C **7**, L450 (1974).

zero over a length $\sim \lambda_{\text{eff}}$. Thus, the flux will be $\sim \pi\lambda_{\text{eff}}^2 H_{c1}$ and equal to Φ_0, or

$$H_{c1} \sim \Phi_0/\pi\lambda_{\text{eff}}^2 \qquad (2-5u)$$

However, if the applied field is just below H_{c2}, the vortices are packed as closely as possible (Fig. 6-1b), and each has an effective area $\sim \pi\xi^2$ or

$$H_{c2} \sim \Phi_0/\pi\xi^2 \qquad (2-5v)$$

These estimates present a useful physical picture and only differ from the correct values by a factor 4 or 2 (Eqs. 5-7c and 5-7d).

H_{c2} and H_{c1} also can be related to the thermodynamic H_c as

$$H_{c2} \approx \sqrt{2}\,\kappa\,H_c$$
$$H_{c1} \approx (H_c/\sqrt{2}\,\kappa)\ln\kappa \qquad (2-5w)$$

thus, $H_{c1}H_{c2} \approx H_c^2 \ln\kappa$

Apart from the $\ln\kappa$ term, H_c is the geometric mean of H_{c2} and H_{c1}. For many compound superconductors, $H_{c2} \gg H_c$, so H_{c1} is very small. H_{c2} vs.

T is shown in Fig. 2-5b for a series of compound superconductors discussed in Section 2-9a. As can be seen, the values are very much larger than the typical 0.1 T values for H_c (Fig. 2-1). Niobium, with $\lambda \sim 350\text{\AA} \; \xi \sim 400\text{\AA}$, is probably the only elemental type II superconductor.

In the vortex state, the vortices act as tiny bar magnets and repel each other. In the absence of vortex pinning centers, the net repulsion energy is minimized when the vortices form a close-packed, hexagonal array, a **vortex lattice**. However, pinning of the vortices often occurs (and is preferable for technological applications); more will be said about this in Chapter 6. Here, we merely remark that the dynamics of the vortices, for conventional as well as high-T_c superconductors, is nontrivial, interesting, and of technological importance. Consider a current **J** flowing perpendicularly to **H** in the vortex state. Then, as a result of the Lorentz force, the vortices are pushed in the **J×H** direction. This motion induces an electric field parallel to **J**, which absorbs energy from the circuit and appears as a resistance!

2-6 BCS Theory

2-6a Introduction — Almost half a century after the discovery of superconductivity, a comprehensive, microscopic theory was proposed by Bardeen, Cooper, and Schrieffer (1957), which is called the BCS theory. This theory gives an excellent account of the equilibrium properties of superconductors and, to quote from their conclusion section, "this quantitative agreement, as well as the fact that we can account for the main features of superconductivity, is convincing evidence that our model is essentially correct." For this work, they received a Nobel Prize in 1972.

The BCS results involve complicated, many-body, quantum-mechanical equations, and a study would take us far afield. BCS is probably best studied by reading the original paper (well worth the time) with the help of some of the superconductivity texts listed in the Bibliography. Fortunately, a complete understanding of BCS is not necessary in order to appreciate many of the qualitative aspects. In this section, we present a few important BCS conclusions, particularly those that can be contrasted to high-T_c results (Chapter 5). See Section 5-2 for more details of the symmetry of the paired state.

2-6b Cooper Pairs and BCS Introduction — Cooper pairs (1956) are the basis of the BCS theory. Cooper considered a non-interacting Fermi gas at 0 K, so all the states are filled for $k \leq k_F$. To this Fermi gas, two electrons are added, which occupy states with $k > k_F$ because of the Pauli-exclusion principle. Then, it is assumed that a net attractive electron-electron interaction exists between these two electrons when their energies are within $\hbar\omega_c$ of the Fermi energy E_F. This attraction occurs because as the first electron moves through the crystal, it distorts (by means of a phonon) the structure in such a manner that a second electron reduces its energy by moving through the distorted structure. For electrons in this energy range, it is suggested that the interaction is attractive and larger than the repulsive, screened-Coulomb interaction and this is called **phonon–mediated pairing**.

The electron-electron interaction, U, scatters a pair of electrons with crystal momenta $(\mathbf{k}, -\mathbf{k})$ to $(\mathbf{k'}, -\mathbf{k'})$. The scattering matrix element is $U_{\mathbf{k}, \mathbf{k'}}$. Cooper (and BCS) assumed that $U_{\mathbf{k}, \mathbf{k'}} = -U_0$ (independent of $\mathbf{k}$, called **weak coupling**) for $\mathbf{k}$ states within a cutoff energy $\hbar\omega_c$ of E_F. The matrix element was taken as zero for energies outside the range $E_F \pm \hbar\omega_c$. Cooper showed that even if both electrons are restricted to having momenta outside the Fermi sphere, they will have a bound state lying below $2E_F$. This bound state is called a **Cooper pair**. Thus, clearly there may be an instability in the electron system if this interaction is operative.

Cooper's work suggested the direction for a microscopic theory. However, it remained for BCS to determine the stable ground state in the presence of pairing. BCS assumed the k-independent, attractive, weak-coupled interaction discussed above. They also assumed a spherical Fermi surface. *With these two simplifying assumptions, they were able to predict universal (material-independent) characteristics of the superconducting state. In fact, it is surprising how well their predictions agree with experiment, considering these simplifying assumptions. However, we should probably be prepared for deviations from their predictions, since real metals do not fit these assumptions very well.*

2-6c BCS Results — We discuss some of the BCS predictions. Let $N(E)$ be the **density of states** at energy E of one-spin orientation, that is, the number of states between energy E and E+dE for electrons of one-spin orientation. (See the Problems). By writing a many-body wave function using the Cooper pair idea, BCS were able to show that there is a second-order phase transition to a new electron state with the transition temperature expressed as

$$k_B T_c = 1.14 \, \hbar\omega_c \, e^{-1/N(E_F)U_0} \qquad (2-6a)$$

where $N(E_F)$ is the density of normal-state electrons at the Fermi energy. The cutoff phonon energy $\hbar\omega_c$ is related to the Debye energy $\hbar\omega_D$, in which

$\omega_D \propto M^{-\frac{1}{2}}$, where M is the atomic mass. Thus, the **isotope effect** comes out of the theory in a natural manner, namely, $T_c \propto M^{-\frac{1}{2}}$, which had been suggested by Fröhlich in 1950 and was found to be in excellent agreement with experiments on mercury isotopes (1950).

BCS found a gap in the allowed states about E_F. Actually, by 1957, many experiments indicated a gap of the order of $k_B T_c$. BCS calculated the 0 K energy gap, $2\Delta(0)$, to be

$$2\Delta(0) = \frac{\hbar\omega_c}{\sinh\left[1/N(E_F)U_0\right]} \approx 4\,\hbar\omega_c\,e^{-1/N(E_F)U_0} \qquad (2-6b)$$

The idea of a gap in the density of states is shown in Fig. 2-6a, in a greatly exaggerated manner. The normal-state, quadratic, free-electron-like density of states is shown, which at 0 K is filled up to E_F. For a superconductor, the 0 K gap is centered at E_F and it "pushes" allowed states into energy regions just below and above the gap, as indicated. (This result is similar to starting with the empty-lattice approximation and turning on a crystal potential.) However, Fig. 2-6a is highly schematic, since typically $E_F \sim 5$ eV and $2\Delta(0) \sim 30$ K ≈ 2.6 meV. The Fermi energy may be taken as the zero of the energy scale leading to the interchangeable use of $N(E_F)$ and $N(0)$ to mean the same thing (i.e., the normal-state density of states at the Fermi energy). The modified superconducting density of states, $N_S(E)$, for $E > \Delta$ and $E < -\Delta$, is

$$N_S(E) = N(0)\frac{E}{(E^2 - \Delta^2)^{\frac{1}{2}}} \qquad (2-6c)$$

For energies within the gap, the density of allowed states is zero. This expression is singular at the edges of the gap ($E = \pm\Delta$), and for positive energies the result is shown in detail in Fig. 2-6b. Note: *the total number of states is unaltered by the interaction.* Those formally in the gap are "pushed out" by the interaction. Thus, the number of superconducting states (Eq. 2-6c) is much larger at T=0 K than just below T_c.

Since the expressions for $k_B T_c$ and $2\Delta(0)$ have the same form, their ratio is independent of the electron density of states and the electron-phonon matrix element, yielding the parameter-free result,

$$2\Delta(0)/k_B T_c = 3.52 \qquad (2-6d)$$

Table 2-2 shows that experimental results for many elemental superconductors are close to 3.5, as are results for many compound superconductors.

By solving the gap equation at elevated temperatures, the temperature dependence of the gap, $2\Delta(T)$, can be calculated. Some results for

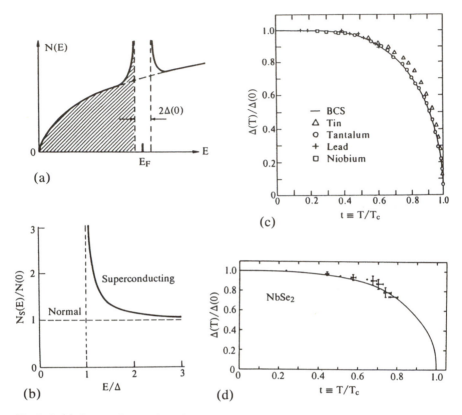

Fig. 2-6 (a) A normal-state, free electron, quadratic density of states is filled with electrons up to E_F. Then, the superconducting energy gap $2\Delta(0)$ is shown and the allowed states that were in the gap are pushed into regions just above and below the gap. (b) More detail of the density of states above the Fermi energy (taken as zero). (c) The BCS reduced energy gap vs. reduced temperature is shown and compared to experimental results for some elemental metals. For details, see P. Townsend and J. Sutton, Phys. Rev. **28**, 591 (1962). (d) The same as the previous figure but for the type II compound $NbSe_2$ measured perpendicular to the **c** axis. For details of the infrared absorption measurements, see B. P. Clayman and R. F. Frindt, Solid State Commun. **9**, 1881 (1971).

elemental metals are shown in Fig. 2-6c, where excellent agreement with BCS can be seen. The gap vanishes at T_c, and just below this value, $\Delta(T)$ can be approximated by

$$\Delta(T) \approx AT_c(1 - T/T_c)^{1/2} \qquad (2-6e)$$

Table 2-3   Of the studied conventional superconductors, the 2H polytypes of NbS_2, $NbSe_2$, andTaS_2 show some strongly anisotropic superconducting properties. Some of the experimental results are listed here. The units for $N(E_F)$ are states/(atom-eV) as in Table 2-2. See Kinoshita's article for references to the original papers.

Mat.	$T_c(K)$	$\lambda_c(\text{Å})$	$\lambda_{ab}(\text{Å})$	$\xi_c(\text{Å})$	$\xi_{ab}(\text{Å})$	$N(E_F)$	λ_{ep}
$NbSe_2$	7.1	4800	1600	37	110	2.2	0.74
TaS_2	0.6	48000	4100	80	960	-	0.41

where $A = 3.06$.

A result for $\Delta(T)$ is shown in Figure 2-6d for the compound super-conductor $NbSe_2$ ($T_c \approx 7$ K). This is in a family of layered materials that have an anisotropic structure, and some of the superconducting properties mirror this anisotropy (Table 2-3). The results in Fig. 2-6d are from infra-red (IR) transition measurements with the IR electric field parallel to the **ab** plane. Thus, the **ab** plane gap (Δ_{ab}) is measured, $2\Delta_{ab}/k_B T_c = 3.7$, and the agreement with the temperature dependence of $\Delta(T)$ is reasonably good. Due to the large anisotropy of some of the superconducting properties (Table 2-3), and because of similar behavior in the high-T_c materials (Chapter 5), an examination of the anisotropy of the gap in this and related anisotropic conventional superconductors would be interesting.

2-6d   Specific Heat — Below T_c, one consequence of a gap in the electronic density of states is that the electronic specific heat is different from that of a normal metal. Recall that for a normal metal, the electronic specific heat can be written as

$$C_{en} = (2/3)\pi^2 N(0) k_B^2 T \equiv \gamma T \qquad (2-6f)$$

where the subscript en refers to the electron component in the normal state. At very low temperatures, the phonons contribute an approximate (Debye) $C_{ph} \approx AT^3$ term that we can ignore in most of our discussion. In fact, the phonon term is smaller than the electronic contribution (Eq. 2-6f) in the conventional superconductor temperature range.

BCS calculated the electronic contribution in the superconducting state, C_{es}. At very low temperatures, where $2\Delta(T)$ is almost independent of temperature, an exponential increase in C_{es} is expected due to thermal excitation of carriers across the gap. C_{es} is shown in Fig. 2-7a along with the, linear in T, C_{en} result. For conventional superconductors, C_{en} is usually obtained from measurements in a magnetic field, with $H > H_c$ or H_{c2}. Most

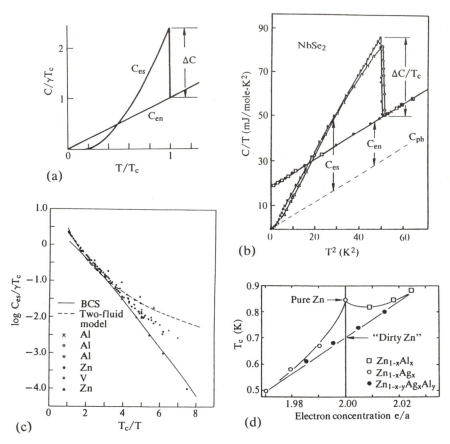

Fig. 2-7 (a) C_{en} and C_{es} vs. reduced temperature showing the exponential rise of C_{es} at low temperature and ΔC at T_c. (b) The normal-electron, superconducting, and phonon contributions to the specific heat are shown for NbSe$_2$ ($T_c \approx 7.2$ K). For details, see P. Garouche, J. J. Veyssie, P. Manuel, and P. Molinie, Solid State Commun. **19**, 455 (1976). (c) log $C_{es}/\gamma T_c$ vs. T_c/T for several elemental superconductors. At low temperatures, much of the experimental data lies above the BCS curve. For details, see H. A. Boorse, Phys. Rev. Letters **2**, 391 (1959). (d) T_c of Zn alloys vs. electron concentration for the systems indicated. The lines are guides to the eye. For details, see Allen and Mitrovic (Notes), p. 62; D. Farrell, J. G. Park, and B. R. Coles, Phys. Rev. Lett. **13**, 328 (1964); P. W. Anderson, Proc. of the 7th Int. Conf. Low Temp. Phys., p. 298 (University of Toronto Press, 1961).

elementary SSP texts show results from an elemental superconductor that are in excellent qualitative agreement with Fig. 2-7a. However, at T_c, BCS found a jump in the specific heat proportional to T_c.

Table 2-4 Data relevant to the specified heat jumps at T_c for some elemental superconductors. See the article by Crow and Ong in Lynn, Ed. (Bib.), Table 7.1, for further references.

Element	T_c (K)	$\Delta C/\gamma T_c$	Θ_D (K)	λ_{ep}
Al	1.16	1.45	428	0.38
Zn	0.85	1.27	309	0.38
Ga	1.08	1.44	325	0.40
Cd	0.52	1.40	209	0.38
In	3.40	1.73	112	0.69
Sn	3.72	1.60	200	0.60
Tl	2.38	1.50	79	0.71
Ta	4.48	1.69	258	0.65
V	5.30	1.49	399	0.60
Pb	7.19	2.71	105	1.12
Hg	4.16	2.37	72	1.00
Nb	9.22	1.87	277	0.82

$$\Delta C \equiv (C_{es} - C_{en})_{T_c} = N(0)(-d\Delta^2/dT)_{T_c} = 9.4N(0)k_B^2 T_c \quad (2-6g)$$
$$\Delta C/C_{en} = \Delta C/\gamma T_c = 1.43$$

The latter expression represents a (universal) jump in the electronic specific heat at T_c. In most of the elemental superconducting metals, $\Delta C/\gamma T_c$ is in good agreement with Eq. 2-6g. However the strong-coupled superconductors (Pb, Hg, and Nb) yield high values, as can be seen in Table 2-4. **Strong-coupled superconductors** (see Section 2-7) are those materials in which the superconducting pairing energy is not negligible compared to the phonon energies. Values of the density of states at the Fermi energy listed in Tables 2-2 and 2-3 are calculated from the γ values obtained from the electronic-specific heat in the normal state (Eq. 2-6f). At very low temperatures ($T_c/T > 10$), the electronic specific heat is approximated by

$$C_{es}/\gamma T_c = 2.37(T_c/T)^{3/2} \exp(-1.76T_c/T) \qquad (2-6h)$$

This result shows that as temperature is increased, C_{es} rises essentially exponentially as carriers are excited across the gap. A simple, useful result, approximately valid for $2.5 < T_c/T < 6$, is $C_{es}/\gamma T_c = 8.5 \exp(-1.44T_c/T)$.

Figure 2-7b shows C_{en} and C_{es} for NbSe$_2$ ($T_c \approx 7.2$ K). If the low-temperature phonon-specific heat is approximated by (C_{ph}) by AT^3, the normal-state specific heat is

$$C = C_{en} + C_{ph} = \gamma T + A T^3 \qquad (2-6i)$$

Thus, a plot of C/T vs. T^2 yields γ as an intercept, and C_{en} and C_{ph} are shown in Fig. 2-7b. NbSe$_2$ is type II (large κ value) and also highly anisotropic (Table 2-3). Yet the results are qualitatively similar to those in Fig. 2-7a.

However, there is more to C_{es} than just its value at T_c. In many conventional superconductors, deviations from the BCS C_{es} result are observed at low temperatures. The log plot in Fig. 2-7c for the reduced C_{es} vs. T_c/T shows that the BCS prediction is close to a straight line, but with a slight downward curvature. This result is much different from the $\propto (T/T_c)^3$ result from the two-fluid model (Eq. 2-2d), which is shown dashed. The data for vanadium (and many other materials) are in excellent agreement with the BCS curve. However, other data lie above the BCS curve at low temperatures. Similar but larger deviations at low temperatures have been observed in Pb, Nb, and other conventional superconductors (Notes). Deviations above the BCS result at low temperatures could be due to the existence of more states at lower energies than those envisioned by BCS, which, in turn, could be due to the coexistence of smaller superconducting gaps, anisotropy in the gap, as well as spin-density waves.

2-6e Anisotropic Superconducting Gap — BCS assumed a spherical Fermi surface and an isotropic phonon-mediated electron-pairing interaction (i.e., independent of the **k** direction). However, Fermi surfaces, even in the simplest metals, are far from spherical. (They have a strong **k** dependence.) Nevertheless, BCS accounts for most of the outstanding properties of conventional superconductors. However, it has been argued that there is experimental evidence for a range of superconducting gaps. For example, some specific heat data (e.g., Fig. 2-7c) can be understood with a range of $2\Delta(0)/k_B T_c$ values.

Calculations of anisotropy in the superconducting gap have generally proceeded by assuming (a) an isotropic phonon spectrum and an anisotropic Fermi surface, (b) an anisotropic phonon spectrum and a spherical Fermi surface, (c) both an anisotropic phonon spectrum and Fermi surface. It has been argued that phonon anisotropy is the principal source of gap anisotropy and, for example, a calculation in lead (Pb) yields ~10% variation of the gap, 2.86 meV and 2.55 meV in the [100] and [110] directions, respectively.

Tunneling experiments in different crystallographic directions in tin yield energy gaps that vary from 4.3 to 3.1$k_B T_c$. Tunneling and ultrasonic attenuation in various directions in single crystal niobium yield similar variations. Specific heat experiments in lead are interpreted in terms of two distinct gaps (4.1 and $1.1k_BT_c$). Specific heat experiments in niobium near T_c show a gap of $3.6k_BT_c$. However, at much lower temperatures, a gap of only one-tenth this size becomes noticeable (Notes).

The results shown in Fig. 2-7d make another case for an anisotropic superconducting gap. The T_c of Zn-based alloys is plotted vs. the electron-to-atom ratio (e/a). In the "dirty" alloys, T_c vs. e/a is linear, while in the "cleaner" alloys, it shows cusplike behavior. Anderson's theory of **dirty superconductors** argues that gap anisotropy will not be affected if the impurity scattering rate $h/\tau_k \ll \Delta$; this scattering rate applies to the clean alloys. However, if $h/\tau_k \gg \Delta$, then the rapid electron scattering causes the electrons to sample the gap in many **k** directions during the time interval h/Δ. This sampling would cause the gap anisotropy to be averaged in dirty superconductors. Note from Fig. 2-7d, a reasonably large $\Delta T_c/T_c \approx 20\%$ is found.

Thus, it seems clear that experiments in conventional (and elemental) superconductors indicate the existence of an anisotropic superconducting gap (i.e., an energy gap that varies with **k** direction). Calculations also show that gap anisotropy should be expected. A few references to some of this work are listed in the Notes.

2-6f   Coherence Effects — The BCS superconducting state is a phase-coherent superposition of occupied one-electron states. (These one-electron states can be treated independently in the normal state.) This phase-coherent superposition can have striking effects when calculating matrix elements in perturbation theory. For certain classes of perturbations, contributions to the transition probabilities have the same sign and add coherently (case I, e.g., ultrasonic attenuation). For other perturbations, terms in the transition probabilities subtract (case II, e.g., nuclear-spin transition probabilities). The difference depends on whether the perturbation is even or odd under time reversal.

Figure 2-8 shows a schematic result for case I and II behavior. Experimental ultrasonic (acoustic phonon) attenuation data is in good agreement with the case I result shown in the figure. Nuclear-spin transition probabilities can be measured in superconductors. The spin-lattice time (T_1) is the inverse of the transition probability and, indeed, it is found that $1/T_1$ follows case II, as expected. Just below T_c, the ratio T_{1n}/T_{1s} (n and s refer to the normal and superconducting states) rises as in Fig. 2-8. This rise is sometimes called the **coherence peak** or the **Hebel-Slichter peak** after the people who first observed this effect in aluminum (Al). For both case I and II, the relaxation rates at much lower temperatures approach zero exponentially, since all of the excitations above the superconducting gap

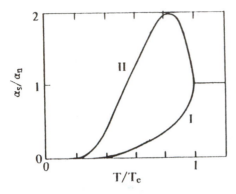

Fig. 2-8 The temperature dependencies of the ratio of the superconducting to normal low frequency absorption processes as predicted by the coherence factors in BCS theory for case I and II.

freeze out. Indeed, exponential freeze-out behavior at low temperatures is experimentally observed for both cases. *The fact that BCS pairing theory can predict (via coherence factors) such remarkably different behavior in a natural manner is one of the triumphs of this theory.*

To calculate $1/T_1$ vs. T to fit experimental data, the superconducting density of states (Fig. 2-6b) must be broadened or the case II coherence factor will go to infinity just below T_c due to the singularity in the BCS density of states. It has been argued that this broadening is consistent with anisotropy in the superconducting gap (Section 2-6e). Experimental evidence for this argument is found in comparing $1/T_1$ measurements in Al to those in Al-Zn and Al-Ge alloys. Compared to pure Al results, the magnitude of the coherence peak increases with alloying, which can be understood in the same manner as the data in Fig. 2-7d. As the material becomes dirty, due to the alloying, the more rapid electron scattering causes a sampling of the gap in many **k** directions. This sampling averages the superconducting gap. From the nuclear spin-lattice relaxation time measurements in Al and these alloys, a ratio of the anisotropy of the gap to the gap is estimated to be about 10%. However, decreasing the magnitude of the coherence peak can also be accounted for by strong-coupled BCS (Section 2-7), since this gives finite lifetimes to the quasi-particle states. Such lifetimes limit the sharpness of density of state peaks. See the Notes for references.

2-7 Strong-Coupled Superconductors

2-7a Introduction — If the electron-phonon coupling is strong (as opposed to weak, as assumed in BCS) then the quasi-particles have a finite lifetime and are thus damped. It has been shown (Notes) that the finite lifetime decreases both $\Delta(0)$ and T_c, but T_c is decreased more because it is determined at finite temperature, and hence in the presence of thermal phonons. Then, one of the effects of strong-coupling is to increase the ratio $2\Delta(0)/k_B T_c$ above 3.52 (Eq. 2-6d). The temperature dependence of the superconducting gap is also modified by the damping effect. Some tunneling results of $\Delta(T)$ show these effects. Although the deviations from BCS may be small, they are in excellent agreement with experiment, as discussed later.

A general indicator of strong-coupling in superconductors, and hence deviations from (weak-coupled) BCS, is the ratio $(T_c/\Theta_D)^2$, where Θ_D is the Debye temperature. For elemental superconductors (Table 2-4), this ratio is the largest in Pb, Hg, and Nb. Hence, the largest effects should be expected for these materials and, indeed, that is the case. In fact, the electron-phonon coupling parameters λ_{ep} (defined below) are larger in these than in the other elements (Table 2-4). This coupling parameter is related to the product of the phonon density of states, $F(\omega)$ and α^2, which describes the electron-phonon interaction. Since λ_{ep} depends on certain details of the crystal structure, we can expect that strong-coupled BCS will not yield universal predictions as does weak-coupled BCS. Thus, the gap-to-T_c ratio, the heat capacity jump at T_c, the temperature dependence of the gap, and other results should depend on the details of the phonon behavior.

2-7b McMillan Equation — Since BCS (1957), there has been considerable progress toward better understanding the electron-phonon interaction in both normal and superconducting metals. Migdal (1958) showed how the strong-coupled electron-phonon interaction could be treated accurately in normal metals. Eliashberg and others extended these ideas to the superconducting state, taking into account the retarded nature of the interaction and treating the damping of the excitations.

Extending these ideas so that comparison to experiments could be made, McMillan calculated the self-energies of normal and paired electrons and used a dimensionless **electron-phonon coupling parameter,**

$$\lambda_{ep} \equiv 2 \int_0^{\omega_{max}} \alpha^2(\omega_q) \, F(\omega_q) \, \frac{d\omega_q}{\omega_q} \qquad (2-7a)$$

where $\alpha^2(\omega_q)$ is the average electron-phonon interaction at frequency ω_q, $F(\omega_q)$ is the phonon density of states, and ω_{max} is the maximum phonon frequency. (Note: this α^2 has nothing to do with the α associated with the isotope effect, Eq. 2-7d and Section 5-5.)

Equation 2-7a should be general and apply for any boson-mediated pairing, not just phonon-mediated pairing. Then $F(\omega)$ would be the boson density of states and $\alpha^2(\omega)$ the coupling strength between the particular bosons and electrons. The ω^{-1} in λ_{ep} increases the importance of the low-frequency bosons with respect to those at higher frequency.

McMillan (1968) numerically solved the finite-temperature, nonlinear Eliashberg equations finding T_c for various classes of strong-coupled superconductors. From these solutions, he constructed an approximate equation that relates T_c to a small number of parameters. One form of the McMillan equation is

$$T_c = \frac{\Theta_D}{1.45} \exp\left[-\frac{1.04(1 + \lambda_{ep})}{\lambda_{ep} - \mu^*(1 + 0.62\lambda_{ep})} \right] \qquad (2-7b)$$

where Θ_D is the Debye temperature and μ^* is the **effective Coulomb-repulsion**. The latter differs from the instantaneous Coulomb repulsion μ by the so-called logarithmic reduction of the Coulomb repulsion due to the screening by the other electrons of the Coulomb repulsion between pairs of electrons. Screening occurs because the Coulomb coupling propagates more rapidly than the phonon coupling. Analysis yields

$$\frac{1}{\mu^*} = \frac{1}{\mu} + \ln\left(\frac{\omega_{el}}{\omega_{ph}} \right) \qquad (2-7c)$$

where ω_{el}/ω_{ph} is a measure of the ratio of the propagation times. Thus, ω_{el} can be taken as the plasma frequency, or the Fermi energy, whereas ω_{ph} corresponds to the high-frequency cutoff of the phonons. Usually, μ^* is between ~ 0 and 0.2, with $\mu^* \approx 0.1$ being the typical value for most superconductors. For $\lambda_{ep} \ll 1$, the weak-coupling limit, $\lambda_{ep} - \mu^*$ plays the role of $N(E_F)U$ in the BCS results (Eq. 2-6b).

The prefactor in Eq. 2-7b can be written in terms of moments of the phonon distribution, but we avoid that for simplicity. This, or any other prefactor, weights high-frequency phonons so as to yield higher T_c values. However, as discussed, the ω^{-1} in Eq. 2-7a weights the lower-frequency phonons. Large λ_{ep} and large Θ_D lead to the highest T_c values.

From Eq. 2-7b, T_c depends on the isotopic mass directly through Θ_D and implicitly through the ω_{max} dependence of the other terms. McMillan found

$$\alpha = \frac{1}{2}\left[1 - (\mu^* \ln \frac{\Theta_D}{1.45T_c})(\frac{1 + 0.62\lambda_{ep}}{1 + \lambda_{ep}})\right] \qquad (2-7d)$$

where α is the "isotope effect" defined from $T_c \propto M^{-\alpha}$ (and should not be confused with the electron-phonon coupling constant). In weak-coupled BCS, we should have $\alpha = \frac{1}{2}$. Neglecting the strong-coupling correction, $(1 + 0.62\lambda_{ep})/(1 + \lambda_{ep})$,

$$\mu^* = (1 - 2\alpha)/(\ln \Theta_D/1.45T_c) \qquad (2 - 7e)$$

Using the measured isotope shift, α values (Table 5-1) for the polyvalent transition metals, an average value of $\mu^* \approx 0.13$ is obtained. From Eq. 2-7e and a related equation for λ_{ep}, McMillan tabulates some of the properties found in Tables 2-2 and 2-4 for the elemental superconductors.

As mentioned, strong-coupled BCS does not yield universal predictions as does (weak-coupled) BCS; details of the phonon distribution affect predictions. We quote a few results for Pb. $2\Delta(0)/k_B T_c = 3.7$ (3.52 for BCS, Eq. 2-6d). The temperature dependence of $\Delta(T)$ is flatter at low temperature and grows steeper near T_c, with $D \approx 4$ (3.06 for BCS, Eq. 2-6e). The jump in specific heat at T_c is 2.6 (1.43 for BCS, Eq. 2-6g). However, the strong-coupled BCS predictions are different for other superconductors. Systematics have been studied of λ_{ep} in some of the conventional super-conductor-alloy systems. It is found that λ_{ep}, and hence T_c, increases due to relative increases in the lower energy part of $\alpha^2 F(\omega)$ (Notes). This is in agreement with Eqs. 2-7a and 2-7b.

2-7c Maximum T_c? — Papers published before 1986 proposed arguments that result in maximum T_c values (Notes). McMillan's proposal stems from a simplification of Eq. 2-7b to

$$T_c \approx <\omega> \exp[-(1 + \lambda_{ep})/\lambda_{ep}] \qquad (2 - 7f)$$

where $<\omega>$ is an average over the phonon density of states. From his expressions,

$$\lambda_{ep} = C/M <\omega^2> \qquad (2 - 7g)$$

where C is fixed for a given class of materials and M is the average mass of the atoms in the material. Thus,

$$T_c = <\omega> \exp[-M<\omega^2>/(C - 1)] \qquad (2 - 7h)$$

We want to increase λ_{ep} to maximize the exponential terms in Eq. 2-7b by decreasing the average phonon frequency. However, in doing this, we decrease the premultiplier. We can find the maximum T_c as a function of $<\omega>$ from Eq. 2-7h and obtain

$$T_c^{max} = (C/2M)^{1/2} e^{-3/2}$$
$$T_c/T_c^{max} = (2/\lambda_{ep})^{1/2} e^{(1/2 - 1/\lambda_{ep})} \qquad (2-7i)$$

which has a broad maximum for $\lambda_{ep} = 2$. Some of the conventional super-conducting systems such as the Pb-Bi alloys, the Nb-like materials, and the Nb_3Sn materials have λ_{ep} values in the 1.5 to 2.5 range (Notes) and T_c values close to T_c^{max}.

In the analysis, it has been assumed that the average phonon frequency can be decreased as desired, but as the phonon frequencies are decreased, the likelihood increases that some phonon mode will become unstable, which could result in the metal transforming to a different crystal structure; certainly, examples of this structural instability have been found.

2-7d Electron–Phonon Parameter Calculations — Calculations of λ_{ep} are difficult because the self-consistent crystal potential must be calculated for the undistorted crystal as well as for the crystal with its atoms displaced to correspond to a particular phonon. Let a phonon of frequency ω have wavevector q on the ν branch, or $\omega_{q\nu}$. Thus, $\lambda_{ep}(q\nu)$ must be calculated and then averaged over all ω and q in the Brillouin zone.

The electron-phonon parameter can be written as

$$\lambda_{ep}(q\nu) = 2N(E_F) \frac{< | g(nk, n'k', q\nu)|^2 >}{\hbar \, \omega_{q\nu}} \qquad (2-7j)$$

where $N(E_F)$ is the electron density of states per spin per atom, evaluated at the Fermi energy. The electron-phonon matrix element is defined as

$$g(nk, n'k', \; q\nu) = \left(\frac{\hbar\Omega_{BZ}}{2M\omega_{q\nu}} \right)^{\frac{1}{2}}$$
$$\times < \psi_{nk}^0 | \varepsilon_{q\nu} \cdot \frac{\delta V}{\delta R} | \psi_{n'k'} > \delta(k - k' - q) \qquad (2-7k)$$

where Ω_{BZ} is the Brillouin zone volume, M is mass of the moving atoms (the phonon mass), $\varepsilon_{q\nu}$ is the phonon polarization vector, ψ_{nk}^0 is a Bloch wave function with wavevector k and band index ν, and $\delta V/\delta R$ is the change of the self-consistent crystal potential caused by the phonon distortion. The latter is calculated by

$$\varepsilon_{q\nu} \cdot \frac{\delta V}{\delta R} = \frac{V_{q\nu} - V_0}{\bar{u}_{q\nu}} \qquad (2-7l)$$

where $\bar{u}_{q\nu}$ is the r.m.s. phonon amplitude, and $V_{q\nu}$ and V_0 are the self-consistent potentials evaluated for the distorted and undistorted potentials, respectively. In practice, calculations of $\lambda_{ep}(q\nu)$ are computer-intensive, and carried out for a few phonons at various $q\nu$ points in the Brillouin zone that are hoped to yield representative values. Then, $\lambda_{ep}(q\nu)$ is averaged over the Brillouin zone.

The phonon line width, $\gamma_{q\nu}$, can also be determined via

$$\gamma_{q\nu} = \pi N(E_F) \, \hbar\omega_{q\nu}^2 \, \lambda_{q\nu} \qquad (2-7m)$$

These line widths should be experimentally accessible by neutron diffraction, IR, or Raman spectroscopies.

At very high pressures, total energy calculations indicate that hydrogen will become metallic with a primitive-hexagonal structure containing two atoms per primitive unit cell. Of course, since hydrogen is very light, its phonons will tend to have very high frequencies; $\Theta_D \approx 2210$ K has been estimated, $\lambda_{ep} \approx 1.5$ has been calculated, and the renormalized Coulomb repulsion has been estimated, $\mu^* \approx 0.1$.

Using these values for μ^*, λ_{ep}, and the calculated $\Theta_D = 2210$ K, from Eq. 2-7b we obtain $T_c = 210$ K. An even slightly larger value (≈ 230 K) is obtained by numerically solving the Eliashberg equations. It will be interesting to see if this phase of hydrogen is obtained at high pressures and then if it is superconducting at such high temperature.

It had often been thought that large electron-phonon parameters only arise from low-frequency phonons. However, this metallic hydrogen calculation and other work show that this is not the case. The resulting large λ_{ep} along with the large Θ_D lead to very large T_c values. It is probably safer to say that at least large-T_c values can be obtained theoretically.

2-8 Tunneling

The specific heat of superconductors beautifully shows the effect of the opening of a gap below T_c. However, tunneling measurements can yield accurate quantitative values for the gap and its temperature dependence. This technique has been applied, with success, to many of the conventional superconductors and is now used for the high-T_c materials. Thus, tunneling is discussed in some detail. Single particle tunneling in superconductors was pioneered by Giaever and it is sometimes referred to as **Giaever tunneling**. For this work, he shared the 1973 Nobel Prize (with Esaki for tunneling in semiconductors and Josephson for electron-pair tunneling, in superconductors, Chapter 6). Remember, the **density of states** at the Fermi energy is usually written as $N(0)$, so energy is measured from E_F.

2–8a   Tunneling Review — Consider two metals separated by an insulator with zero voltage applied (Fig. 2-9a). If the insulator is thin (of the order of an electron mean free path), then there is a finite probability that an electron from one metal will tunnel through the barrier (insulator) and enter the second metal. Of course, this implies that empty states are available into which the electrons tunnel. Figure 2-9a is drawn for T=0 K; thus, no tunneling is allowed for V=0. However, for T>0 K, electrons in the tail of the Fermi-Dirac distributions may tunnel from one metal into the other.

We should appreciate that the energy-band diagrams in Fig. 2-9 are for the so-called **semiconductor model** of a superconductor. This means that, for the normal metal, a plot of energy (vertical) vs. the density of states is given with the states filled to E_F at T=0 K. For higher temperatures, there is a smearing of the number of occupied states within $\sim k_B T$ of E_F, which is described by the Fermi-Dirac distribution function. This is the usual situation for a metal. The superconductor is represented just like a semiconductor, with an energy gap corresponding to the one-electron superconducting gap (Eq. 2-6b) centered at E_F. For an applied potential difference V, the energy levels of one metal are shifted vertically with respect to the other metal by an amount eV. Tunneling transitions are all horizontal (at constant energy). With the help of the semiconductor model, calculations of contributions to the single-particle currents are straightforward. However, this model is inadequate for dealing with processes in which *electron pairs* play a role, such as Josephson tunneling. This is because the one-electron semiconductor model does not take into account the ground state of paired electrons.

Using the one-electron, semiconductor model, the tunneling current from metal 1 to 2 can be calculated by noting that $N_1 f$ and $N_2(1 - f)$ are just the number of occupied (initial) states in metal 1 and empty (final) states in metal 2, respectively, where f is the Fermi-Dirac function. Then the current is

$$I_{1 \to 2} = A \int_{-\infty}^{\infty} |B|^2 N_1(E)\, f(E)\, N_2(E + eV)\, [1 - f(E + eV)]\, dE \quad (2 - 8a)$$

where V is the applied voltage, N(E) is the appropriate density of states, A is a constant of proportionality, and B is a tunneling-matrix element. If the

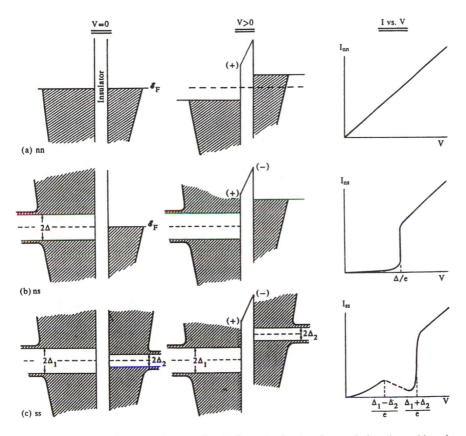

Fig. 2-9 Schematic diagrams of energy (vertical) vs. the density of states in junctions with and without an applied voltage, V. The corresponding I vs. V curves are on the right. The notations n and s refer to normal and superconducting metals. The density of states diagrams are for T=0 K, but the I vs. V results are drawn for slightly high temperatures, which allow a finite current below eV= Δ and $\Delta_1 + \Delta_2$.

reverse current is subtracted, then a general expression for the total current is

$$I = A|B|^2 \int_{-\infty}^{\infty} N_1(E) \, N_2(E + eV) \, [f(E) - f(E + eV)] \, dE \qquad (2-8b)$$

Normal–normal tunneling can be calculated using Eq. 2-8b. The subscript n denotes a normal metal, so

$$I_{nn} = A|B|^2 N_1(0) N_2(0) \int_{-\infty}^{\infty} [f(E) - f(E + eV)] dE$$

$$= A|B|^2 N_1(0) N_2(0) eV \equiv G_{nn}V \tag{2 - 8c}$$

where G_{nn} is the **conductance**. Of course, Eq. 2-8c is Ohm's law, $I \propto V$ (Fig. 2-9a). Note that a temperature-independent conductance is obtained.

Normal-superconducting tunneling also can be calculated using Eq. 2-8b (s denotes the superconductor) and the result is

$$I_{ns} = A|B|^2 N_2(0) \int_{-\infty}^{\infty} N_{1s}(E) [f(E) - f(E + eV)] dE$$

$$= \frac{G_{nn}}{e} \int_{-\infty}^{\infty} \frac{N_{1s}(E)}{N_1(0)} [f(E) - f(E + eV)] dE \tag{2 - 8d}$$

To evaluate this expression as a function of applied voltage (and T), the BCS density of states is required (Eq. 2-6c), and numerical results are easily obtained. However, qualitatively certain limits are straightforward. At $T=0$ K, $I_{ns}(E)$ is zero until $E = \Delta$, so there is no tunneling current until $eV = \Delta$; this result is sketched in Fig. 2-10a, and in Fig. 2-10b, the normalized **differential conductance** is shown at $T = 0$ K. The latter can be straightforwardly obtained via Eq. 2-8d.

$$G_{ns} = \frac{dI_{ns}}{dV} = G_{nn} \int_{-\infty}^{\infty} \frac{N_{1s}(E)}{N_1(0)} \left[\frac{-\partial f(E + eV)}{\partial(eV)} \right] dE \tag{2 - 8e}$$

$$(G_{ns})_{T = 0} = (dI_{ns}/dV)_{T = 0} = G_{nn} [N_{1s}(e|V|)/N_1(0)]$$

where the result at 0 K is also given. Note how the 0 K conductance is proportional to the density of states (Fig. 2-10b). At finite temperatures, the conductance also yields the density of states but smeared by k_BT due to the Fermi-Dirac function. The $T > 0$ K results are also shown in Figs. 2-10a and 2-10b.

In the same manner, tunneling between two superconductors (I_{ss}) can be calculated from Eq. 2-8b. We shall not do this here (see the Notes for references); however, Fig. 2-9c shows the results, which can be appreciated qualitatively. At $T=0$ K, no current flows until $eV = \Delta_1 + \Delta_2$. For $T > 0$ K, the conductance shows a peak at a voltage corresponding to the difference of the superconducting gaps.

2-8b   Tunneling Experiments — Experimental results for a superconductor-insulator-superconductor tunnel junction are shown in Fig. 2-11a for a lead-lead oxide-lead (Pb-I-Pb) junction. $T_c = 7.2$ K and

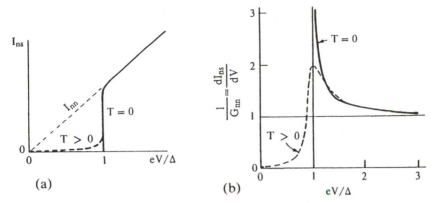

Fig. 2-10 Characteristics of a metal-superconductor tunnel junction for T=0 K (solid lines) and finite temperature (dashed lines). (a) I vs. V. (b) Normalized conductance vs. normalized voltage.

measurements were made for 1.9 K $\leq$ T $\leq$ T$_c$. Note that for T near T$_c$, there is a large amount of current for eV<2Δ; nevertheless, 2Δ(T) can be determined. The resulting temperature dependence of the superconducting gap is shown in Fig. 2-11b along with the BCS calculation. As can be seen, agreement between experiment and BCS is good.

In spite of the good agreement between experiment and the theory (Fig. 2-11b), it can be seen that for t>0.76 (t$\equiv$T/T$_c$), the experimental points tend to lie below BCS, with the reverse happening for t<0.76. These deviations can be understood in terms of strong-coupled BCS theory (Section 2-7). For these superconductors, the strong electron-phonon interaction gives rise to considerable damping of the quasi-particle excitation. Using the strong-coupled BCS theory, numerical calculations can be performed, and these are shown as open circles in Fig. 2-11b (labeled SSW). The agreement with the experimental results is excellent.

2-8c Phonon Structure — In early tunneling measurements in lead, Giaever et al. noticed structure at energies corresponding to the Pb phonons. Figure 2-12a shows the normalized conductance of a good Pb-I-Pb junction below T$_c$, and measured to higher voltages than shown in Fig. 2-11a. These data can be understood by realizing that for **strong-coupled BCS** materials, the superconducting density of states should be

$$N_s(E) = N(0) \; \text{Re} \frac{E}{[E^2 - \Delta^2(E)]^{\frac{1}{2}}} \qquad (2-8f)$$

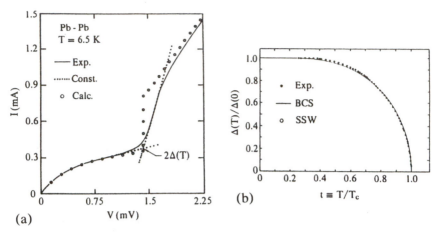

Fig. 2-11 (a) I vs. V for a Pb-I-Pb junction with a construction showing how the superconducting gap is obtained. The open circles are calculated using the BCS density of states. (b) The experimental gap ratio compared to BCS values and strong-coupling calculations (SSW). For details and an explanation of SSW, see F. F. Gasparovic, B. N. Taylor, and R. E. Eck, Solid State Commun. **4**, 59 (1966).

where Re means the real part of the expression that follows. If Δ is real and independent of energy, then this equation reduces to the ordinary (weak-coupled) BCS result (Eq. 2-6c); however, in strong-coupling theory, Δ is complex and energy-dependent. The imaginary part (Im) of Δ corresponds to damping of the superconducting quasi-particles via the creation of phonons. Thus, Im $\Delta(E)$ peaks at energies near the energy of the phonons that are important in the phonon-mediated pairing.

It turns out that it is possible to invert the tunneling data (Fig. 2-12a) and find the energy dependence of α^2F, the product of the electron-phonon constant and the phonon density of states (Section 2-7). The resultant α^2F vs. E for Pb is shown in Fig. 2-12b. The two peaks correspond to peaks in the phonon density of states due to the transverse- and longitudinal-acoustic phonon branches. Since Pb has the fcc crystal structure (one atom per primitive unit cell), there are no optical phonon branches. However, similar results for α^2F vs. E have been obtained for materials that have two or more atoms in a primitive unit cell and thus have optical phonon branches. These results include tin and mercury, where good agreement with the phonon data is obtained.

For crystals at low energy, $\alpha^2F(\omega) \propto \omega^2$, which results from $\omega \propto k$ for phonons in the acoustic branch. If the same material is made amorphous

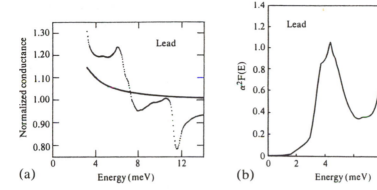

Fig. 2-12 The dots are the experimental normalized (to the normal state) conductance vs. eV for a Pb-I-Pb junction below T_c. The solid lines are calculated from the BCS density of states with no phonon structure. (b) α^2F vs. E for Pb obtained by fitting the data in (a).

(e.g., by condensation on a cold substrate), the density of phonon states at very low energies is increased, causing $\alpha^2F(\omega) \propto \omega^n$, where n can be unity or smaller. Since the ω^{-1} term in Eq. 2-7a emphasizes the low-energy contributions, λ_{ep} should increase, which is experimentally observed. For example Pb-Bi alloys have λ_{ep} in the range of 2, which results from the large atomic masses causing $F(\omega)$ to peak at low energies. However, amorphous $Pb_{0.45}Bi_{0.55}$ has an even larger value, $\lambda_{ep}=2.6$ (with $\mu^*=0.12$).

For conventional superconductors, λ_{ep} may also increase if the material is close to a structural phase transition with soft (i.e., low-energy) phonon modes at a Brillouin zone boundary. These low-energy modes increase the density of phonon states at low energies, increasing λ_{ep} (Eq. 2-7a). The A15 superconductors have some of these phonon instabilities. In high-temperature superconductors, the role of soft-phonon modes is considered (Section 5-6b).

Perhaps the most important result of these tunneling experiments, and analysis using the strong-coupled theory, is that the results provide strong verification that the electron pairing in these superconductors is phonon-mediated. Of course, this statement applies to conventional superconductors and it can be argued that this pairing mechanism may not hold for certain classes of conventional superconductors. These may (or may not) include the d-band metals such as Zr, Ru, and Os (Table 5-1), which have a very small isotope effect. Also, the heavy-electron superconductors (the last three entries in Table 2-1) probably do not have phonon-mediated electron pairing (Section 2-9c).

The tunneling experiments also provide values of α^2F vs. E, which are important for other calculations in superconductors. For example, the electron-phonon coupling parameter (λ_{ep}) can be obtained (Eq. 2-7a). The phonon density of states can be measured by neutron diffraction; thus, the frequency-dependent electron-phonon coefficient (α^2) itself can be obtained. Last, these experiments provide a good test for the strong-coupling BCS theory.

2-9 Other Topics

Besides the conventional superconductors that appear to be well understood in terms of BCS, there are classes of pre-1986 superconductors that exhibit unusual and interesting properties. Some of these materials are mentioned here.

2-9a Magnetic Superconductors — Superconductivity can be destroyed by an external magnetic field (Figs. 2-1 and 2-5). Thus, we might expect that it could also be destroyed if the magnetic movements of the ions in a crystal magnetically order (ferro- or antiferromagnetically) or even if paramagnetic atoms are alloyed into the crystal. A Cooper pair consists of two electrons with equal and opposite crystal momentum and opposite spin, which thus forms a time-reversed pair of one-electron states. An external (or internal) magnetic field breaks this time-reversal symmetry by flipping one spin of the Cooper pair. Indeed, for small concentrations, alloying a superconductor with nonmagnetic impurities has little or no effect on T_c; these impurities are important only in reducing the electron mean free path (Sections 2-4a and 2-6e). However, alloying a superconductor with paramagnetic-impurity rare-earth (RE) atoms in small concentrations ($\sim 1\%$) severely depresses T_c. Further, the reduction of T_c is correlated with the spin of the RE ions, rather than with their magnetic moments (1959). This result leads to a description of the effect of paramagnetic impurities in terms of an exchange Hamiltonian,

$$H_{ex} = N^{-1} \, \Sigma_i \, J_{ex} \, s \cdot S_i \qquad\qquad (2-9a)$$

where J_{ex} is the exchange constant, s the conduction electron spin, S_i the spin of the ith localized atom, and N is the total number of atoms in the sample.

Gapless Superconductivity — Abrikosov and Gor'kov (1961) considered the effect of this spin-exchange Hamiltonian (Eq. 2-9a) for the case of dilute magnetic impurities and obtained a scattering time for a spin-flip process. This process results in a finite lifetime for the Cooper pairs, which reduces T_c. (Scattering from nonmagnetic impurities does not reverse spins, and hence is time-reversal invariant.) From these considerations, they found a rather interesting result. For small concentrations, x, of magnetic impurity, T_c vs. x varies linearly with a slope proportional to J_{ex}. At larger x, T_c decreased more rapidly, becoming zero at a critical concentration x_c, where $x_c \sim 1$ percent. The remarkable feature is that the gap decreases more rapidly with x, becoming zero at $\approx 0.91 x_c$. This is called **gapless superconductivity** and gives some indication of the interesting interplay between superconductivity and magnetism.

Reentrant Superconductors — Discovery of superconductivity in the rare-earth (RE), **Chevrel-phase** compounds $REMo_6X_8$ (X=S or Se) and the XRh_4B_4 (X=Y, Th, or RE) type compounds (1975-1977) has allowed the interplay between superconductivity and magnetism to be studied more easily. Let T_M be the temperature of the onset of cooperative magnetism.

As an example, consider $HoMo_6S_8$, which becomes superconducting at $T_{c1} \approx 1.8$ K, but at a lower temperature, $T_{c2} \approx 0.7$ K, it reenters the normal state, hence, the name **reentrant superconductor**. For this material, $T_{c2} \sim T_M$, where a first-order transition to a normal-state, ferromagnetic metal occurs. Figure 2-13a indicates these results, as part of the $(Ho_{1-x}Eu_x)Mo_6S_8$ phase diagram.

As a second example, $ErRh_4B_4$ is found to be superconducting with $T_{c1} \approx 8.7$ K, then has a reentrant transition $T_{c2} \sim 1$ K $\sim T_M$ (Fig. 2-13b). The figure also shows the $(Eu_{1-x}Ho_x)Rh_4B_4$ phase diagram to be discussed.

Extensive neutron diffraction experiments in the vicinity of T_M have been carried out on both $ErRh_4B_4$ and $HoMo_6S_8$. In the former, just above T_{c2}, there is strong evidence for the coexistence of superconducting regions with ferromagnetic regions. The RE magnetic moments are sinusoidally, spatially modulated with a wavelength of the order of a few hundred angstroms, which is related to the superconducting penetration depth. At lower temperatures, the ferromagnetic energy dominates and pure ferromagnetism occurs, which completely destroys the superconductivity. Similar effects may occur in $HoMo_6S_8$, but the situation is less clear.

Results for the pseudo-ternary system $(Ho_{1-x}Eu_x)Mo_6S_8$ are shown in Fig. 2-13a. The Ho material is a reentrant superconductor. The Eu material has spin-glass behavior and is antiferromagnetic at lower temperatures, although single crystals become superconducting above 10 K under high pressures. In the middle region of this phase diagram, it can be seen that superconductivity occurs at $T_{c1} \sim 2$ K and $T_M \sim 0.5$ K, but below T_M,

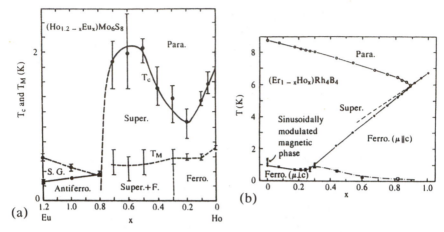

Fig. 2-13 (a) Low-temperature phase diagram of $(Ho_{1.2-x}Eu_x)Mo_6S_8$; early work, incorrectly, used this formula. The correct formula is $(Ho_{1-x}Eu_x)Mo_6S_8$, as used in the text. Reentrant superconductivity is seen on the Ho side with spin-glass and antiferromagnetic behavior on the Eu side. For details, see M. Ishikawa, O. Fischer, and J. Muller in Maple and Fischer, Ed. (Notes). (b) The low-temperature phase diagrams of $(Er_{1-x}Ho_x)Rh_4B_4$. For details, see M. B. Maple, H. C. Hamaker, and L. D. Woolf in Maple and Fischer, Ed. (Bib.).

superconductivity coexists with some sort of magnetic order. The precise nature of the magnetic state below T_M is not clear; however, it appears to have a net zero-field magnetization. Therefore, the material shows the coexistence of superconductivity and ferromagnetic order.

In some of these reentrant superconductors, there may be a tendency for a physical separation between the superconducting electrons and the magnetic electrons. For example, in $RERh_4B_4$, the electron bands that give rise to superconductivity are believed to be composed primarily of 4d-Rh electrons, while the magnetic electron bands are composed primarily of 4f-RE electrons. The physical separation of the superconducting and magnetic electrons may give rise to negligible exchange constants (Eq. 2-9a). Then, only the relatively weak dipole interaction remains to break Cooper pairs.

Superconducting crystals also may have magnetic interactions that lead to an ordered antiferromagnetic phase. In this case, the magnetic ordering produces no macroscopic magnetic field and, on the order of the scale of the superconducting coherence length, the antiferromagnetic exchange field may average to zero. Thus, the coexistence of superconductivity and antiferromagnetism is not surprising. In fact, for some superconductors, below the magnetic ordering temperature, the antiferromagnetism seems only to effect H_{c2} values of the superconductor. In $ErMo_6S_8$ and $SmRh_4B_4$, below the antiferromagnetic ordering temperature, H_{c2} increases, while in $GdMo_6S_8$ and $NdRh_4B_4$, it decreases. Also, some of the heavy-electron metals (Section 2-9c) show the coexistence of superconductivity and magnetic ordering.

This field is rich and many aspects may be related to high-T_c superconductors. The Notes should be consulted for further reading.

2-9b Earlier Oxide Superconductors — The cuprate, high-T_c superconductors were not the first oxides to be found that display superconductivity. The oxygen-deficient perovskite $SrTiO_{3-x}$ was the first (1965) of many (Fig. 5-1). Before 1986, $Ba(Pb_{1-x}Bi_x)O_3$ had the highest T_c of the oxides ($T_c \approx 13$ K). However, the discovery of the cuprates has led to more work in this area, and the highest T_c record (≈ 30 K) now belongs to $(K_{0.4}Ba_{0.6})BiO_3$ (1988), which also has the perovskite crystal structure. It is not clear if these types of oxides have any particular relationship to the superconducting cuprates. Thus, we reserve the term **high-T_c superconductors** only for the cuprates (independent of the value of T_c), since it is felt that the Cu-O planes (Chapter 1) are the distinctive characteristic of these materials. We continue to define **conventional superconductors** as all pre- and post-1986 superconductors that do not contain Cu-O planes.

2-9c Heavy-Electron Metals — The last three compounds listed in Table 2-1 are heavy-electron metals. Generally, heavy-electron metals can be characterized by a low-temperature specific heat that is two or three orders of magnitude larger than ordinary metals; they also have other anomalous properties, and usually are f-electron metals. The normal-state electronic specific heat (Eq. 2-6f) is $C_{en} \propto N(E_F) \propto m^{*3/2}$, where $N(E_F)$ is the density of states at the Fermi energy and m^* is the electron effective mass. Thus, these metals have huge effective masses, hence, the name "heavy." Heavy-electron metals have three known ground states. Some are superconductors (Table 2-1), some order magnetically at low temperatures (e.g., $NpBe_{13}$, U_2Sn_{17}, and UCd_{11}), while some show neither superconductivity nor magnetism (e.g., $CeAl_3$, $CeCu_6$, and UAl_2).

It is becoming clear that at least some, and perhaps all, of the heavy-electron superconductors do not have singlet spin-state, isotropic, s-wave orbital-state pairing, as considered by BCS. Rather, they have singlet spin-state, but d-wave pairing (or perhaps triplet spin-state, p-wave pairing, see Section 5-2b). One of the consequences of d-wave (or p-wave) pairing; is that the superconducting gap goes to zero at certain k-points or lines on the Fermi surface. This leads to the expectation that many of the equilibrium-superconducting properties will be different than those predicted by BCS; for example, see Fig. 5-10b. Actually, due to the large f-electron spin-orbit coupling and the crystal symmetry terms, the classification of the paired electrons in terms of s, d, ... pairing becomes less exact because these terms mix the various orbital pair-states. For these superconductors, it is unlikely that the pairing is phonon-mediated, and it is probably magnetic in origin. Magnon-mediated pairing tends to yield d-wave pairing, but much work remains to be done in this field.

2-9d Organic Superconductors — The first organic superconductors were discovered in 1980, with $T_c < 1$ K. Most organic superconductors are charge transfer salts in which there is a charge transfer between an organic cation molecule and an inorganic anion complex. The π orbitals in the cation are responsible for the metallic conductivity.

An important organic building block in these superconductors is bis(ethylenedithia)tetrathiafulvalene, which is usually referred to as ET. In 1983, $(ET)_2ReO_4$ was found to have a $T_c = 2.5$ K, which was felt to be quite high at that time. It is now known that $(ET)_2Cu[N(CN)_2]Br$ has $T_c = 11.6$ K, rather high by pre-1986 standards.

These charge-transfer organic superconductors, like the cuprate superconductors, are highly anisotropic. Some have appreciable conductivity along only one direction. Many of the ET superconductors have appreciable conductivity in a plane, and for some of these materials, Fermi surfaces have been measured in the conducting plane. An inverse isotope effect has been found for some of the organic superconductors upon replacing hydrogen with deutrium.

For ordinary metals, in the optical spectral region, the inelastic scattering rate for electrons is found to vary with the square of the frequency or the temperature. This is expected for a Fermi liquid. However, for the organic superconductors, the optical data imply that the inelastic scattering rate varies linearly with both the frequency and the temperature. The same linearity is found for the high-T_c superconductors. (Section 5-4a). Both organic and high-T_c superconductors have low electron densities and good electrical conductivities in fewer than three dimensions. These similarities have helped to generate recent increased interest in the organic superconductors.

2-9e ^{3}**He** — Both ^{4}He and ^{3}He liquids have a phase transition at ~2.2 K and ~ 3×10^{-3} K, respectively. Below these temperatures, the liquids are said to be in a super-fluid phase which has remarkable properties, hence, the word "super." However, the physics of these two phases is completely different.

A helium atom has two electrons in the 1s orbital state with spins antiparallel. Thus, the total electron angular momentum is zero, as is the total angular momentum of the ^{4}He nucleus (two protons and two neutrons). Hence, ^{4}He is a boson and can condense into a phase (Bose condensation) with an unlimited occupation of particles in the ground state. On the other hand ^{3}He is a fermion, since it has a nuclear spin I=1/2 (two protons and one neutron) and obeys Fermi statistics. Thus, Bose condensation is not allowed in ^{3}He.

Soon after the BCS-pairing explanation of superconductivity, there was speculation of pairing in other systems and liquid ^{3}He was considered. Theoretically, it was felt that the strong repulsive core of the electron cloud was not compatible with s-wave pairing as in BCS. Rather, a higher orbital angular momentum might be required for the necessary attractive interaction. Indeed, it appears that the two phase transitions in the 2 to 3 mK range in liquid ^{3}He result from the formation of a BCS-like paired, spin-triplet state (Section 5-2) with p orbital angular momentum. References in the Notes can be consulted for details of this condensed-fermion state.

Problems

1. Free Electron Gas — (a) The **density of states** $N(E)$ is the number of single-electron states, per unit energy range, for one-spin orientation. For a free electron gas, show that the density of states is

$$N(E) = 2(2^{1/2}V/\pi^2)(m/\hbar^2)^{3/2} E^{1/2}$$

(b) Show that the low-temperature electron specific heat is

$$C_V = 2(\pi^2/3)k_B^2 N(E_F) T$$

Show that $N(E_F) = 3N/E_F = 3N/k_B T_F$, so the specific heat per mole is

$$C_V = \pi^2 z N_A (T/T_F) \equiv \delta T$$

where N_A is Avogadro's number and each atom contributes z electrons to the gas.

2. Critical current of a type I superconductor — Consider a long cylindrical superconducting wire of radius R.   (a) Show that the field in the wire produced by a current I is

$$H(r) = \frac{2I}{cR} \; \frac{r}{R}$$

where c is the velocity of light. (b) The critical current can be defined as the current that yields H_c (at the surface).   Thus, show that $I_c = H_c cR/2$.

3. Two-fluid model — Using the free energy given in the two-fluid model (Section 2-2), show that the fraction of "normal" electrons is $(T/T_c)^4$. Using standard thermodynamic relations, determine the electronic heat capacity in the superconducting phase.

4. Little–Parks experiment demonstrates that the **fluxoid** is quantized rather than the flux. Describe how this is done. (See W. A. Little and R. D. Parks, Phys. Rev. Letters **9**, 9 (1962) and Phys. Rev. **133**, A97 (1964), and M. Tinkman, Phys. Rev. **129**, 2413 (1963)).

Chapter 3

Structures

Nevertheless, let us take this business seriously and
spare no pains; success is never automatic in this world.

Herodotus, "The Histories"

Compared to structures encountered in most areas of solid-state physics, those of the high-T_c crystals are complicated. The existence and stacking of the Cu-O planes is critical, since it is electrical conduction in those planes that gives rise to superconductivity. In a broad sense, the important aspects of the Cu-O planes are covered in Chapter 1 and pictured in Fig. 1-1. There are now many high-T_c Cu-O planar superconductors. Most of them fall into families of structures with different numbers of immediately adjacent Cu-O planes. (We let n be the number of immediately adjacent planes.) For $n>1$ **immediately adjacent Cu-O planes**, each Cu-O plane is separated by a sparsely populated plane of Y or Ca atoms. Immediately adjacent sets of n Cu-O planes are in turn separated from the next set of n Cu-O planes by metal-O **isolation planes** (also called **charge reservoirs**), where the metal atoms usually are La, Ba, Tl, or Bi (Fig. 1-1).

3-1 Overview

A single Cu-O plane is shown in Fig. 1-1a, where the square-planar bonding of Cu to four O atoms can be seen. Figure 1-1b shows the Cu-O planar arrangement in $(La_{2-x}Sr_x)CuO_4$. In this crystal, as well as in all of the high-T_c materials, the Cu-O distance is rather short, ≈ 1.90Å. However, the Cu-O planes are relatively far apart, ≈ 6.6Å, with two La-O planes between the Cu-O planes. Figure 1-1c shows the skeleton structure for $YBa_2Cu_3O_{7-\delta}$. For this material, there are two immediately adjacent

Cu-O planes (≈ 3.2Å apart), and these two Cu-O planes (n=2) are >8.2Å from the next two Cu-O planes. The three metal-O isolation planes that separate the two immediately adjacent Cu-O planes in Y123 are indicated by lightly dashed planes in Fig. 1-1c. For a high-T_c crystal like $Tl_2Sr_2Ca_2Cu_3O_{10}$, there are three (n=3) Cu-O immediately adjacent planes (separated by two Ca planes), again separated from each other by ≈ 3.2Å. The three Cu-O planes are separated from the next sets of three immediately adjacent Cu-O planes by four metal-O isolation planes (two Tl-O and two Sr-O) so that the sets of three Cu-O planes are ≈ 11.6Å apart.

Most of the high-T_c structures are tetragonal or nearly tetragonal (having a small orthorhombic or other distortion). For the purposes of our discussion, we can consider the materials as tetragonal. This simplification has the important advantage that the families of well-studied structures can then be classified according to just two tetragonal space groups (symmetries), and it is then easy to make comparisons among the materials. The n Cu-O planes are always perpendicular to the **c** axis; that is, they are parallel to the **ab** plane (Fig. 1-1).

Table 3-1 is a list of materials in the more widely studied families along with descriptive abbreviations; we use the ones in the fourth column. The idealized chemical formulae are given, the approximate T_c values, and the n values. (Remember that n refers to the n Cu-O planes that are immediately adjacent to each other but always separated by a sparsely populated Y or Ca plane for $n \geq 2$.)

Consider the abbreviations 1-Tl(n) and 2-Tl(n) as examples. The n Cu-O immediately adjacent planes are separated by two Sr-O planes plus one Tl-O plane for 1-Tl(n), and by two Sr-O, plus two Tl-O planes for 2-Tl. Figures 3-1, 3-2, and 3-3 show some of the often studied structures for n=1, 2, and 3. The Cu-O planes are fairly obvious in these figures. Depending on n and the structure, above and below almost all of the Cu atoms there may be an oxygen atom (called O_z or the apical O atom). O_z atoms are actually part of the metal-O isolation planes in between the immediately adjacent n Cu-O planes. This is clear from the distances; in the Cu-O plane, the Cu-O distance is typically 1.9Å. However, the Cu-O_z distances are >2.4Å, which places these atoms relatively far away from each other.

The larger n, the higher the T_c values, at least up to $n \geq 3$. For n=1, 2, and 3, values of T_c lie in the approximate ranges 0-85 K, 60-92 K, and 105-125 K, respectively, although examples outside these broad ranges can be found. Some of the 1-Tl(n) and 2-Tl(n) materials with n>3 have been synthesized, but often the T_c values have not increased, although note the exception for 1-Tl(n=4). This could be due to a saturation of the n vs. T_c,

Table 3-1 A partial list of the high-temperature superconductors. The ideal chemical formulae are listed; the T_c values tend to be the higher ones reported; n is the number of the immediately adjacent Cu-O planes in the unit cell, which is also the number of Cu-O planes per primitive cell. Last, some other notations that appear in the literature are given. $(Ba,K)BiO_3$ is added for completeness.

Formula	T_c (K)	n	Notations	
$(La_{2-x}Sr_x)CuO_4$	38	1	La(n=1)	214
$(La_{2-x}Sr_x)CaCu_2O_6$	60	2	La(n=2)	--
$Tl_2Ba_2CuO_6$	0-80	1	2-Tl(n=1)	Tl2201
$Tl_2Ba_2CaCu_2O_8$	108	2	2-Tl(n=2)	Tl2212
$Tl_2Ba_2Ca_2Cu_3O_{10}$	125	3	2-Tl(n=3)	Tl2223
$Bi_2Sr_2CuO_6$	0-20	1	2-Bi(n=1)	Bi2201
$Bi_2Sr_2CaCu_2O_8$	85	2	2-Bi(n=2)	Bi2212
$Bi_2Sr_2Ca_2Cu_3O_{10}$	110	3	2-Bi(n=3)	Bi2223
$(Nd_{2-x}Ce_x)CuO_4$	30	1	Nd(n=1)	T'
$YBa_2Cu_3O_7$	92	2	Y123	YBCO
$YBa_2Cu_4O_8$	80	2	Y124	--
$Y_2Ba_4Cu_7O_{14}$	40	2	Y247	--
$TlBa_2CuO_5$	0-50	1	1-Tl(n=1)	Tl1201
$TlBa_2CaCu_2O_7$	80	2	1-Tl(n=2)	Tl1212
$TlBa_2Ca_2Cu_3O_9$	110	3	1-Tl(n=3)	Tl1223
$TlBa_2Ca_3Cu_4O_{11}$	122	4	1-Tl(n=4)	Tl1234
$CaCuO_2$	--	1	n=∞	--
$(Nd,Ce,Sr)CuO_4$	30	1	-	T*
$(Ba_{0.6}K_{0.4})BiO_3$	30	-	-	BKBO

or it could be due to improper doping or other stoichiometry problems with higher n materials. In principle, there is no reason why materials of the type 3-Tl(n), 4-Tl(n), and so forth can not be made. The same might apply to the Bi materials. If the trend found for 1-Tl(n) to 2-Tl(n) is maintained, then higher T_c values might result. More work remains to be done in these areas.

The Cu-O planes dominate the high-T_c materials from the electrical and superconducting points of view and, to a large extent, from the structural point of view. Thus, we should expect highly anisotropic normal-state properties, and probably anisotropic superconducting properties as well. These anisotropies will become apparent in the following chapters.

This completes the overview of the cuprate high-T_c structures. For a first reading, no further structural details are required, figures for particular structures will be referenced in the later chapters as needed. The rest of this chapter discusses these and some related structures in more detail. If desired, the reader may go to the next chapter.

3-2 Structures

We shall discuss the different families of high-T_c structures starting with the one discovered first, La(n=1). This crystal has a body-centered tetragonal (bct) lattice (two lattice points per conventional unit cell). Then we discuss the 2-Tl(n) and 2-Bi(n) families, both of which have the same bct lattice and thus have many similarities with La(n=1). Following these, the 1-Tl(n) crystals are discussed, which have a primitive tetragonal (pt) lattice; Y123 is a special case of 1-Tl(n=2). Actually, Y123 is structurally complicated because it has a superconducting Y123-O_7 form and an insulating Y123-O_6 form. Other high-T_c structures are then considered.

This paragraph is a reminder of some basic aspects of lattices and structures. Remember that a **lattice** is just an arrangement of points in space (no atoms). One unit cell of a bct lattice can be seen in Fig. 3-1a (or Fig. 3-1b) by imagining that the Cu atoms are lattice points and all of the atoms are ignored. Two unit cells of a pt lattice can be seen in Figs. 3-2a, 3-2b, or 3-2c by imagining that the Tl atoms are the lattice points and all of the atoms are ignored. A **structure** is a lattice with a group of atoms (a basis) attached (convoluted) to each lattice point. Thus, *the basic translational periodicity comes from the lattice* and the bct and pt lattices have different Brillouin zones (Fig. 4-8).

In describing and understanding these complicated structures, there is a nomenclature problem. In Y123, for example, Cu atoms are found in Cu-O planes as well as in Cu-O chains, and there are three (or five, depending on subtlety) different kinds of O atoms. Each of these different types of Cu (or O) atoms have different Wyckoff symbols and crystallographers use these, as they are well-defined, universally agreed upon, and contain important symmetry information. In the physics and chemistry literature, these symbols along with other notations are used. In many non-structural high-T_c papers, an arbitrary numerical notation, O(1), O(2), ..., O(5) for Y123 is used. With any arbitrary notation, the problem

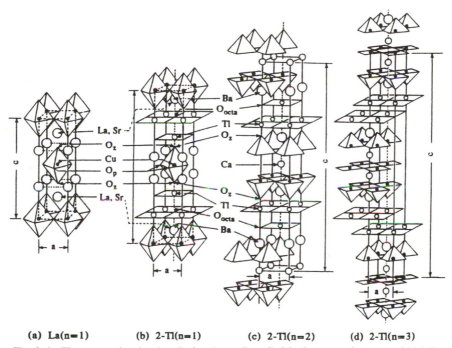

(a) La(n=1)   **(b) 2-Tl(n=1)**   **(c) 2-Tl(n=2)**   **(d) 2-Tl(n=3)**

Fig. 3-1 The conventional unit cells for the well-studied body-centered tetragonal high-T_c materials. The space group is I4/mmm(D_{4h} [17]). Oxygen atoms are not shown explicitly, but are located at the intersection of the straight lines.

is that it is arbitrary, no helpful information is apparent, and it is often different in different papers. In this book, we use a **descriptive notation** where oxygens in the plane are O_p, along the chain in Y123 are O_c. To either of these oxygen atom symbols, we may add a comment if it is important to note that they are along the **b** or **a** axis in orthorhombic Y123; thus, $O_p(b)$ or $O_c(b)$ are along the **b** axis in the plane and chain, respectively; if along the **a** axis, $O_p(a)$ or $O_c(a)$. Oxygen atoms in isolation planes of the Tl superconductors are labeled O_{octa} because they are octahedrally coordinated. Last, the apical O atoms in these materials are labeled O_z because they are along the z axis, directly above or below the Cu_p atoms; perhaps they might have been labeled O_{ap} for apical. Hopefully, this descriptive notation will be helpful.

3-2a   La(n=1) — From Fig. 3-1a, we can see how the La(n=1) structure can be described. The Cu-O planes, perpendicular to the **c** axis, are on mirror planes. Above (and below) are planes of La-O, then another La-O plane shifted by $(\frac{1}{2}, \frac{1}{2}, 0)$, causing the body centering, followed by the next Cu-O plane. The La-O planes (the isolation planes) are not flat but corrugated, and they seem to have a secondary role as far as superconductivity is concerned. There are two formula units (La_2CuO_4) in the conventional bct unit cell, as can be seen Fig. 3-1a. Thus, the primitive unit cell (not shown) contains just one formula unit. The translational symmetry is that of a bct lattice, which, of course, determines the Brillouin zone (Fig. 4-8).

The structure of La($n=1$) shown in Fig. 3-1a is tetragonal and has been known for many years as the K_2NiF_4 structure, and, in the high-T_c field, it is also called the **T structure**. At high temperatures (depending on x as shown in Fig. 4-4c) there is a transition to an orthorhombic phase, which has structural parameters that differ by small amounts from those in the tetragonal phase. This phase transition is discussed in Section 3-3.

For $x=0$ ($La_{2-x}Sr_x$)CuO_4 is an insulator and is typically hole-doped with Sr^{2+} replacing some of the La^{3+}. The holes reside principally on the Cu-O plane and result in metallic behavior for $x \gtrsim 0.05$ (Section 4-7). From most diffraction experiments, it is reported that Sr substitutes randomly for the La atoms. However, ordering of Sr atoms on the La sites has been recently reported from single-crystal x-ray structural studies (Notes). Several types of partial ordering have been observed with the loss of some oxygen. It has been suggested that this single-crystal ordering may be the reason that ceramics of this material generally have higher T_c values and a larger superconducting volume-fraction than single crystals with the same Sr content.

Hole doping of La($n=1$) with Na and K atoms, ($La_{2-x}K_x$)$CuO_{4-\delta}$, has also been reported. However, the situation is more complicated than Sr doping because of the resulting La and O vacancies. More work is required on the processing steps. Nevertheless, T_c values in the same range as Sr doping have been found, with the samples remaining tetragonal.

3-2b   2-Tl(n) — The general formula is $Tl_2Ba_2Ca_{n-1}Cu_nO_{4+2n}$ (Table 3-1). These body-centered tetragonal crystals (Figs. 3-1b to 3-1d) consist of n immediately adjacent Cu-O planes, with a Ca plane between each immediately adjacent Cu-O plane for $n \geq 2$. That is, two Cu-O planes have one plane of Ca atoms, three Cu-O planes have two planes of Ca atoms, and so forth as shown in the figures and chemical formulae. Separating these immediately adjacent Cu-O planes is a (corrugated) Ba-O plane, two

Tl-O planes, and another Ba-O plane before the next Cu-O plane is encountered. The large·distance ($>11\text{Å}$) between the n Cu-O planes is due to these four Ba-O and Tl-O isolation planes.

Three different types of oxygen atoms can be seen in these structures. Oxygen atoms in the Cu-O plane are labeled O_p; each Cu atom has four O_p atoms as nearest neighbors (Figs. 3-1 and 1-1a). Oxygens directly above and below (along the c axis) the Cu atoms are called O_z or the apical oxygens; their nearest neighbors are one Cu atom, four Ba atoms, and one Tl atom; they are part of the Ba-O plane perpendicular to the c axis. The oxygen atoms that are part of the Tl-O planes are labeled O_{octa} because they are octahedrally surrounded by six heavy-metal atoms ($5\text{Tl} + 1\text{Ba}$). Note La(n=1) contains O_p and O_z atoms, but clearly not O_{octa}.

As in the La(n=1) case, for all of the bct 2-Tl(n) crystals, there are two formula units per conventional bct unit cell (Fig. 3-1) and one formula unit per primitive unit cell. It is worthwhile to focus, for example, on the Ba atoms in Fig. 3-1b to convince yourself that the two Ba atoms near the top and bottom of the figure have the identical environment as those toward the middle of the unit cell.

3-2c 2-Bi(n) — In general, 2-Bi(n) crystals (Table 3-1) have the same structure as do 2-Tl(n) crystals. The principal differences are that Ba is usually replaced by Sr in the 2-Bi(n) crystals. Also, the Bi materials usually have an orthorhombic distortion.

3-2d 1-Tl(n) — The general formula for these materials is $TlBa_2Ca_{n-1}Cu_nO_{3+2n}$ and *two* unit cells for the n=1, 2, 3 structures are shown in Fig. 3-2 for these primitive tetragonal (pt) crystals. Note there is one formula unit in each primitive unit cell for these 1-Tl(n) crystals. Two unit cells are shown in Fig. 3-2 to ease comparison with the bct unit cells in Fig. 3-1 for 2-Tl(n). The structures consist of n immediately adjacent Cu-O planes, with a plane of Ca between the Cu-O planes for $n \geq 2$. For this structure, there is only one Tl-O (uncorrugated) plane, rather than two as in the 2-Tl(n) family; this fact is included in the notation. Thus, separating the n immediately adjacent Cu-O planes is a Ba-O plane, one Tl-O plane, and another Ba-O plane. Bi analogues for these compounds have not been reported yet. For 1-Tl(n=1), the distance between Cu-O planes is just the c-axis unit cell length = 9.7Å.

For the 1-Tl(n) structures, note that all three types of oxygen atoms (O_p, O_z, and O_{octa}) occur. For each oxygen-atom type, the environment in

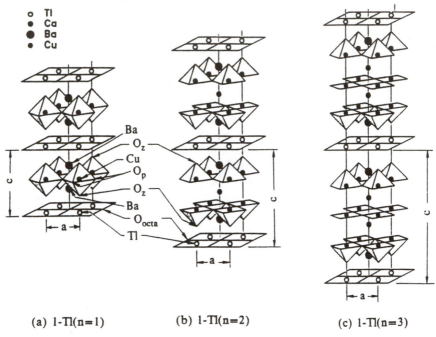

o Tl
• Ca
● Ba
• Cu

Ba
O_z
Cu
O_p
O_z
Ba
O_{octa}
Tl

(a) 1-Tl(n=1)   (b) 1-Tl(n=2)   (c) 1-Tl(n=3)

Fig. 3-2 (a) to (c) are the structures of the tetragonal 1-Tl(n=1, 2, 3) high-T_c crystals with space group P4/mmm(D_{4h}[1]). In each diagram, two unit cells are shown to enable easier comparison to the body-centered tetragonal unit cells in Fig. 3-1. Oxygen atoms are not shown explicitly, but are located at the intersection of the straight lines.

1-Tl(n) and 2-Tl(n), as well as La(n=1) and Y123, is the same, so the nomenclature is generally useful.

3-2e Distances — Although the bct and pt, 2-Tl(n) and 1=Tl(n), crystal structure families are very different, the interatomic distances are all very similar. The distances between the planar-Cu atom and O_p in La(n=1), 2-Tl(n=1), and 1-Tl(n=1) are just one half of the **a** direction unit cell length. Thus, these Cu-O_p distances are 1.89, 1.93, and 1.93Å, respectively. For the n≥2 structures, a small amount of corrugation (puckering) of these planes results in these interatomic distances being a few percent larger. However, all of these materials have Cu-O_p distances in the 1.89 to 1.94Å range. On the other hand, the Cu-O_z distances are all ≈2.41Å. Thus, Figs. 3-1 and 3-2 can be misleading; Cu-O_p corresponds to a short, strong covalent bond very much different from the weak Cu-O_z

bond. To talk of the apical oxygen (O_z) as part of a pyramid or, in La(n=1), as part of a Cu-O octahedra is probably not useful. From the interatomic distances, it is clear that the Cu atoms are covalently, square-planar bonded to four O_p atoms (Fig. 1-1a). This strong, square-planar Cu-O bonding affects the structural, normal-state, and superconducting properties.

3-2f Y123 — $YBa_2Cu_3O_{7-\delta}$ (or Y123-$O_{7-\delta}$) is complicated because of its variable oxygen content and two different types of four-coordinated Cu. Independent of δ, Y123 has n=2 with two closely spaced (3.2Å) Cu-O planes separated by a Y plane. These two immediately adjacent Cu-O planes are ≈8.2Å from the neighboring sets of Cu-O planes.

Y123-O_7 (Fig. 3-3) can be considered as a special case of 1-Tl(n=2) (Fig. 3-2b), with Y replacing Ca, Cu replacing Tl, and O_{octa} moved so that it is directly between two Cu atoms and is now labeled O_c(b), as shown in Fig. 3-3a. There are now two different types of Cu atoms, so they are labeled Cu_p for the Cu in the two Cu-O planes, and Cu_c for the atoms that can be thought of as replacing Tl and are now on a four-coordinated chain, as will be described. O_p and O_z atoms still occur in this structure (Fig. 3-3a), but the oxygen atoms in the chain are labeled O_c, while in 1-Tl(n=2), they were O_{octa} atoms. The occurrence of O_c causes this material to be orthorhombic, with the Cu_c-O_c-Cu_c axis arbitrarily taken as the **b** axis (as opposed to the **a** axis). Thus, we could completely specify this chain-oxygen atom as O_c(b) to distinguish it from a largely unoccupied position O_c(a), which is indicated by a cross in Fig. 3-3a. Similarly, if useful, the planar oxygens along the **b** and **a** axes could be labeled O_p(b) and O_p(a).

With this useful labeling behind us, Fig. 3-3b emphasizes the Cu coordination. The distances between Cu_p-O_p(b) and Cu_p-O_p(a) are 1.96 and 1.93Å, respectively, very similar to those found for the Cu-O planes in the 2-Tl(n), 2-Bi(n), and 1-Tl(n) materials discussed previously. The Cu_p-O_z distance is 2.30Å, which is very large; this same large distance is also found in the other high-T_c superconductors. Now consider the Cu chains. The Cu_c-O_c(b) distance is 1.94Å (half of the **b** axis unit cell length) and the Cu_c-O_z distance is 1.84Å. *Thus, like the Cu_p atoms, the Cu_c atoms are strongly covalently bonded to four O atoms in a square-planar configuration.* However, the Cu-O planes extend indefinitely in two directions (the **ab** plane), while the Cu-O chains extend indefinitely in only one direction (the **b**-axis direction).

Due to the intense interest in Y123, there have been many complete determinations of the atomic positions in this structure. As an example of the detail, joint x-ray and neutron refinements of the Y123 structure have

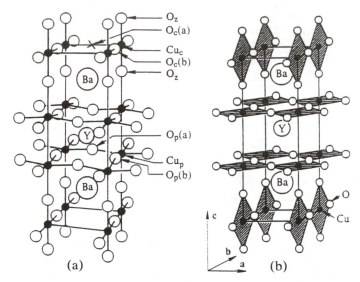

Fig. 3-3 Representations of the orthorhombic Y123, Y123-O$_7$, structure. (a) emphasizes the labeling of the atom positions. A normally unoccupied position, O$_C$(a), is indicated for discussion purposes. (b) emphasizes the Cu coordination polyhedra.

been reported. Since x-rays are less sensitive to the oxygen atoms (because the scattering depends on the electron density) simultaneous refinement of data from both techniques can lead to more accurate results. For example, the *anisotropic* thermal parameters for *all of the atoms* can be obtained. Specialized techniques such as pulsed-neutron diffraction have also been applied to Y123 and some of these references are given in the Notes. From these types of measurements from 10 K to 300 K in Y123-O$_7$, it appears that the interatomic distances and anisotropic thermal factor behave smoothly through T$_c$.

For about one year(1987), Y123 was the only crystal with T$_c$>77 K, so it was felt that the Cu-O chains may play a significant role. However, with the discovery of the other high-T$_c$ families, it became clear that the chains cannot be of paramount importance. Similar to the Ba-O and Tl-O planes, the chains appear to act only as charge reservoirs as well as structural elements.

Y123-O$_6$ is trivial to understand in terms of the Y123-O$_7$ structure shown in Fig. 3-3. The one fewer oxygen atom in the formula is obtained by removing the O$_C$(b) atom from the Y123-O$_7$ structure. Thus, for Y123-O$_6$, the Cu-O planes remain intact, as do all the other atoms, except the chains now contain linear, two-coordinated Cu atoms in O_z-Cu_c-O_z "sticks," which results in Y123-O_6 being tetragonal, but with similar unit cell dimensions as Y123-O_7.

Note, for both Y123-O_7 and Y123-O_6, there is only one formula unit per unit cell as in all of the pt 1-Tl(n) materials (Fig. 3-2). However, Y123-O_7 is actually orthorhombic as mentioned.

Considerable effort has been extended to determine the structure(s) of Y123 with intermediate oxygen content (between 7 and 6). Most of the neutron and x-ray diffraction measurements report unit cell lengths and positional parameters that vary continuously with δ, although rather rapid variations occur near the metal-insulator phase transition ($\sim$Y123-$O_{6.35}$); however, no new structures are found. On the other hand, electron diffraction measurements, which are more sensitive to such subtleties, support O_c vacancy-ordered structures for intermediate oxygen-content Y123.

3-2g Other High-T_c Structures — We discuss some other structures that are of interest in the high-T_c field.

T′ Structure — The bct structure of Nd_2CuO_4, is shown in Fig. 3-4a and it is called the T′ structure, or Nd(n=1). It has a close relationship to the La(n=1) structure, which is also called the **T structure** and is shown again in Fig. 3-4b. Both structures have single isolated Cu-O planes perpendicular to the **c** axis, and in both structures the La or Nd atoms are along the **c** axis directly above and below the Cu atoms. The difference between these structures is the arrangement of the additional O atoms, O(4d) and O_z(4e), respectively (where the 4c and 4d are Wyckoff symbols). With the proper doping, both materials are high-T_c superconductors. What is interesting is that it appears that the T structure can only be easily hole-doped, while the T′ structure can only be easily electron-doped; for example, $(Nd_{2-x}Ce_x)CuO_4$ is electron-doped because Ce^{4+} replaces Nd^{3+}. The reason for the relation between the structure and type of doping is not presently known.

T* Structure — The T* structure is shown in Fig. 3-4c, which is related to both the T′ and T structures. If, using Figs. 3-4a and 3-4b, the lower half of the T′ unit cell is combined with the upper half of the T unit cell, a picture of the (primitive-tetragonal) unit cell of the T* structure is obtained, and it is shown in Fig. 3-4c (note that n=1). The T* structure has Cu-O planes perpendicular to the **c** axis and can be found in a range of solid solutions of the type, $(La,Nd,Ce,Sr)_2CuO_4$. For example, one type is $(Nd_{2-x-y}Ce_ySr_x)Cu_2O_4$, of which a particular example is $(Nd_{1.3}Ce_{0.3}Sr_{0.4})Cu_2O_4$. For both the T′ and T* materials, T_c values in the

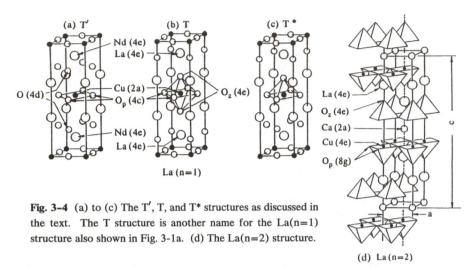

Fig. 3-4 (a) to (c) The T′, T, and T* structures as discussed in the text. The T structure is another name for the La(n=1) structure also shown in Fig. 3-1a. (d) The La(n=2) structure.

20-30 K range have been obtained. In all three of these structures, the distance between the single Cu-O planes is ≈ 6.6Å, while the Cu_p-O_p distances are the usual ~1.9Å.

La(n=2) — If another Cu-O plane is added to La_2CuO_4, and therefore a Ca plane to separate the two immediately adjacent Cu-O planes, then $La_2CaCu_2O_6$ is obtained. The structure is shown in Fig. 3-4d, and we call it La(n=2), since it is the n=2 extension of La(n=1). For La(n=2), $T_c \approx 60$ K has been recently reported (Table 3-1, Notes). The discovery of superconductivity in La(n=2) is a good example of the importance of proper doping. There had been many papers on La(n=2) showing the structure and methods of preparation. However, superconductivity was not obtained until it was properly doped, which was achieved by getting the oxygen content close to 6.0 by annealing under a high pressure of oxygen (Notes).

La(n=3) would be $La_2Ca_2Cu_3O_8$, but it has not yet been reported to be superconducting. There are isostructural insulating crystals with one, two, and three planes, namely, Sr_2TiO_4, $Sr_3Ti_2O_7$, and $Sr_4Ti_3O_{10}$. Another structural analogy of La(n=2) is $(Ba,Sr)SrInO_6$, which has two closely spaced In-O planes separated by a plane of Sr atoms. In principle, for the T′ structure, n=2, 3, ... , Cu-O plane materials might be synthesized. Hence, we use the label Nd(n=1). However, these structures have not yet been reported.

Y123 related structures — A projection along the a axis of Y123-O_7 is shown in Fig. 3-5a. The corner-sharing chains directed along the **b**

axis can be easily seen. The Cu-O planes, which project onto parallel lines, are also observable.

Growth at higher oxygen pressure can yield the compound, $YBa_2Cu_4O_8$ (Y124), shown in Fig. 3-5b. This high-T_c material ($T_c \sim 80$ K) has double edge-sharing chains along the **b** axis as well as the two immediately adjacent Cu-O planes, as in Y123. This compound appears to have less of a tendency to lose oxygen than Y123. By replacing some of the Y with Ca atoms (changing the doping), T_c can be increased to ~ 90 K.

Growth at intermediate oxygen pressures can yield crystals of $Y_2Ba_4Cu_7O_{14+x}$ (Y247), whose structure is shown in Fig. 3-5c, and $T_c \sim 40$ K. It is related to both Y123 and Y124 in a manner similar to the way T* is related to the T and T′ structures. Y247 consists of the usual two immediately adjacent Cu-O planes, but the chains alternate between single corner-shared ones and double edge-shared ones. In the single corner-shared chains, there is considerable disorder between $O_c(b)$ and $O_c(a)$ type sites as indicated by the x in the figure. In a sense, Y247=Y123+Y124.

Both Y124 and Y247 have **a** and **b** lengths about the same as those in Y123, but the **c** axes are longer to accommodate the extra chains. Nevertheless, the interatomic distances are very similar for all three of these crystals.

n = infinity structure — By considering 2-Tl(n) or 1-Tl(n) and letting n go to infinity, the n=∞ material, $CaCuO_2$, is found (Fig. 3-6a). This tetragonal material consists of nothing but Cu-O planes with Ca atoms between them. Another conceptual way to obtain this structure is to start with a cubic perovskite (Fig. 3-6b) and remove the planes of O atoms (one O per unit cell), as indicated in going from Figs. 3-5b to 3-5a.

$CaCuO_2$ does not exist as such; rather, some Sr atoms are required to stabilize the n=∞ structure. For example, $(Ca_{0.86}Sr_{0.14})CuO_2$ forms this structure. However, this material is not a metal but an insulator, and attempts to dope it have not succeeded as yet. However, see the Notes.

$BaBiO_3$-type superconductors — We have defined high-T_c superconductors as those with Cu-O planes. Thus, $BaBiO_3$-type superconductors are not considered as "high-T_c superconductors." Also, high-T_c superconductors have many highly anisotropic properties and are magnetic in their insulating phases; $BaBiO_3$-type superconductors have neither of these properties. In 1975, it was discovered that $Ba(Bi_{1-x}Pb_x)O_3$ was superconducting over a large range of x with a maximum $T_c = 13$ K for x=0.75. More recently, $(Bi_{0.6}K_{0.4})BiO_3$ was found to have $T_c \approx 30$ K. Although this is not high by present standards, it certainly is high by pre-1986 standards, and thus the material is worth mentioning.

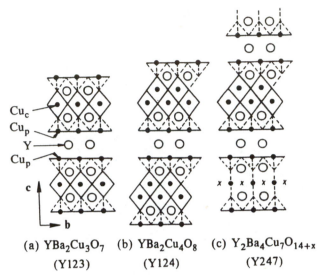

Cu_c
Cu_p
Y
Cu_p

c

b

(a) $YBa_2Cu_3O_7$ (b) $YBa_2Cu_4O_8$ (c) $Y_2Ba_4Cu_7O_{14+x}$

(Y123)   (Y124)   (Y247)

Fig. 3-5 (a) The (100) projection of Y123. (b) and (c) (100) projections of the two structures as labeled, both of which are closely related to Y123. See the text for details.

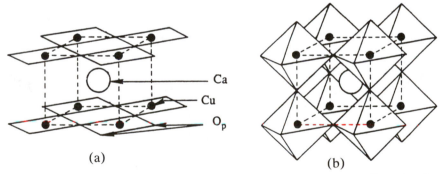

Ca
Cu
O_p

(a)   (b)

Fig. 3-6 (a) The tetragonal $CaCuO_2$ or $n=\infty$ structure. The oxygen atoms are located at the intersections of the straight lines. (b) The cubic perovskite ABO_3 structure with space group $Pm\bar{3}m(O_h^1)$. Undistorted $BaBiO_3$ has this structure.

$BaBiO_3$ has an almost undistorted ABO_3, cubic **perovskite structure** (Fig. 3-6b). Each Bi atom (B site) is octahedrally coordinated by six O atoms. When undoped, it is not a metal. However, when doped on the A site with potassium, it becomes metallic, superconducting, and it appears to remain strictly cubic to 0 K. However, in light of the slightly distorted behavior of pure and Pb-doped $BaBiO_3$, EXAFS of the local symmetry of the K-doped superconductor might prove interesting.

3-3 Other Phases

3-3a   La(n=1) — The high-T_c superconductor $(La_{2-x}Sr_x)CuO_4$, has one **structural phase transition** for x between 0 and about 0.3. It has the high-temperature tetragonal (HTT) phase, then a low-temperature orthorhombic (LTO) phase. $(La_{2-x}Ba_x)CuO_4$ also has this structural transition, but for $0.1 \lesssim x \lesssim 0.15$, it has another phase transition to a low-temperature tetragonal (LTT) phase, then

$$HTT \rightarrow LTO \rightarrow LTT$$

These structures and phase transitions are isomorphic to those found in La_2NiO_4. Next, we describe these three structures and their relationships.

For ease of description, we use La_2NiO_4 as the prototype for these three structures. The HTT phase is exactly that shown in Fig. 3-1a (and Fig. 3-4b), with Ni replacing the Cu atoms. The structure has a bct lattice with space group I4/mmm − D_{4h}^{17} in the International crystallographic and Schoenflies notations, respectively, and there are two formula units in the cell (Z=2). Instead of the conventional cell shown in Fig. 3-1a, we show the identical structure with an equivalent face-centered tetragonal (fct) unit cell (Z=4) in Fig. 3-7c. The **a** and **b** axes of the fct cell are rotated by 45° and are $\sqrt{2}$ longer than those of the bct cell. (Thus, the fct cell has twice the volume and number of atoms.) Also, the crystallographic space group notation is F4/mmm for the fct cell rather than I4/mmm for the bct cell, but since they are merely different cells for the *identical structure*, the Schoenflies symbol (D_{4h}^{17}) is the same, as indicated in the figure. The atoms are numbered for later use. We use this larger fct cell because it has the same number of atoms (and essentially the same dimensions) as the unit cells of the LTO and LTT unit cells.

The LTO structure (Fig. 3-7b) is obtained by rotating the NiO_6 octahedra about the [010] axis such that the rotation direction alternates between nearest neighbors in the basal plane, as required by corner sharing of the NiO_6 octahedra. Thus, in the (001) planes, neighboring octahedra are counter-rotated. However, in the (010) planes, neighboring octahedra are rotated in the same direction. Thus, the arrows in Fig. 3-7b represent actual displacements of the oxygen-planar atoms, which then result in corrugated (puckered) Ni-O planes. The apical oxygen atoms (O_z) and La atoms, along the **c** axis above and below the Ni atoms, displace in the **ab**

(a) HTT   (b) LTO   (c) LTT

F4/mmm(D_{4h} [17])   Abma(D_{2h} [18]) $P4_2$/ncm(D_{4h} [16])

≡I4/mmm(D_{4h} [17])

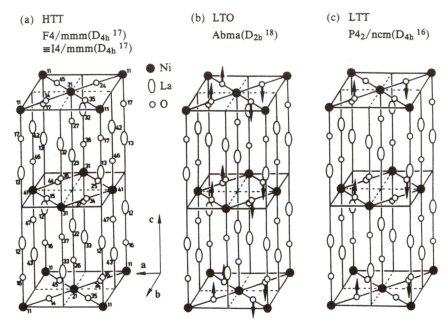

Fig. 3-7 (a) The undistorted high-temperature tetragonal (HTT) structure of La_2NiO_4. The numbers label the atoms. (b) and (c) The oxygen displacements for the Ni-O plane are shown in the low-temperature orthorhombic (LTO) and low-temperature tetragonal (LTT) structures, respectively. The apical oxygen and La atoms are also displaced, but these displacements are omitted for clarity; see the text. For details, see R. Geick and K. Strobel, J. Phys. C. **10**, 4221 (1977); Burns and Glazer (Bib.); and G. Burns, F. H. Dacol, D. J. Buttrey, D. E. Rice, and M. K. Crawford, Phys. Rev. B **42**, 10777 (1990).

direction as would be expected for approximately rigid NiO_6 octahedra. For clarity, however, the displacement vectors for La and O_z atoms are not shown. In the LTO structure, the rotation about the corner atom (11) and the A-face centered atom (41) is the same, but this rotation is opposite to that of the other two face-centered atoms (21 and 31). Thus, the space group is A-centered. If the **a** and **b** axes are interchanged, the resulting space group is Bmab(D_{2h}^{18}).

The LTT structure is shown in Fig. 3-7c. The displacements are obtained by rotating the NiO_6 octahedra about <110> directions such that, in the basal plane, nearest neighbors are counter-rotated, as required by corner sharing of the NiO_6 octahedra. Along the **c** axis, the rotation axis alternates between [110] and [1$\bar{1}$0] direction. The resulting planar-oxygen atom displacements are indicated by arrows. As can be seen, half of the O_p atoms are undisplaced. The O_z and La atoms displace in the **ab** direction as expected for approximately rigid NiO_6 octahedra; however, displacement vectors are not shown for clarity.

As mentioned, superconductivity occurs in La(n=1) with Sr doping in the LTO phase, and the LTT phase is not observed. Also, superconductivity occurs in La(n=1) with Ba doping in the LTO phase. However, for Ba doping in the approximate range $0.1 \lesssim x \lesssim 0.15$, the LTT phase is observed below ~70 K, and superconductivity appears to be suppressed in this LTT structure. However, in some closely related materials with the LTT structure, superconductivity has been observed (e.g., $La_{1.6}Nd_{0.25}Sr_{0.15}CuO_4$).

Neutron-phonon-dispersion measurements (phonon energy vs. **k**, Section 5-6d) of single-crystal La_2CuO_4 have been performed at high temperatures. Classical soft-mode behavior (i.e., a phonon energy approaching zero energy, or zero restoring force) was observed for a zone-boundary ($\frac{1}{2}$, $\frac{1}{2}$, 0) phonon, which is just the phonon describing the Cu-O plane tilting for the HTT to LTO transition. A frozen phonon calculation for this tilting indeed results in a double-well potential for this mode, indicating a structural instability in the HTT phase against such motion and a phase transition at some high temperature.

3-3b Y123 — A phase transition in Y123 is occasionally mentioned, since it changes from orthorhombic to tetragonal as the temperature is raised to ~500°C. Of course, in this temperature range, the oxygen content of the sample is decreasing. Thus, this change of structure is not a phase transition in the usual sense, since it does not happen at constant composition.

In fully oxygenated Y123, there also have been occasional reports of a phase transition at very low temperatures. This transition is due to ordering of the $O_c(b)$ atoms to form a zigzag chain and thus causing the unit cell to double along the **b** direction. However, such claims do not appear to be substantiated.

3-3c Y123 with Intermediate Oxygen Content — For fully oxygenated Y123, the plane containing the Cu-chain atoms (Cu_c) is shown in Fig. 3-8a and the structure is labeled **Ortho I** for reasons that will be apparent. The Cu_c-O_c chains along the **b** axis are apparent and the repeat length along **a** is about 4.0Å.

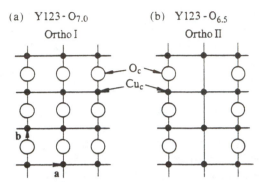

(a) Y123-$O_{7.0}$ (b) Y123-$O_{6.5}$

Ortho I   Ortho II

Fig. 3-8 (a) The plane containing Cu_c and O_c of Y123-O_7. The O_c positions are fully occupied and the structure has been called Ortho I. (b) The same plane but for the proposed Ortho II structure of Y123-$O_{6.5}$, where every other O_c row is missing.

There is considerable interest in Y123 with oxygen content intermediate between O_7 and O_6, what can be called Y123-O_x. Electron diffraction studies, on carefully prepared samples with intermediate oxygen content, reveal a variety of superstructures with different repeats along the **a** axis. Probably the most important superstructure is observed for an oxygen range from approximately Y123-$O_{6.61}$ to Y123-$O_{6.28}$, which extends from superconducting material on the T_c=60 K plateau (Fig. 4-5b) to tetragonal, insulating material, respectively. A model for this superstructure is the so-called **Ortho II structure**, which, for Y123-$O_{6.5}$, is shown in Fig. 3-8b. In this structure, every other oxygen row along the **b** axis is missing, causing a doubling of the repeat along the **a** axis. There is a tendency to associate the Ortho II structure with the T_c=60 K plateau, but this is problematic, since this structure is observed over such a wide oxygen content.

3-3d Other Distortions — We have been taking the simplest point of view in discussing the structures. That is, we have assumed that all but the Y123 materials are tetragonal except for the phase transitions discussed in Section 3-3a. Here, brief mention is made as to which structures actually are tetragonal, which have slight orthorhombic distortions from tetragonal symmetry, which have incommensurate modulations, and related matters. It is important to realize that some of the distortions and vacancy ordering observed in the high-T_c materials are only observed by electron diffraction in TEM. This is particularly true for the distortions in the Tl-based superconductors and the vacancy ordering in the Y123 materials. Thus, the amount of charge density involved is very small and electron diffraction is required to detect the distortion and orderings. Also, TEM observations are always slightly suspect because the observation is on only a few small grains from a huge batch of polycrystalline matter. Relatedly, care must be exercised that the high-energy electron beam (~100 keV) does not induce transitions or otherwise modify the structure of the very minute crystallites studied.

La(n=1) has been discussed in Section 3-3a. La(n=2) appears to remain tetragonal down to 4 K for the dopings studied. The T' and T* materials (Table 3-1) also appear to remain tetragonal to 4 K.

All of the 1-Tl(n) and 2-Tl(n=2 and 3) are reported to be tetragonal. However, 2-Tl(n=1) appears as tetragonal and pseudo-tetragonal (orthorhombic) modification, depending on growth conditions.

All of the 2-Bi(n) materials are pseudo-tetragonal, really orthorhombic with **a** and **b** axes 45° and $\sqrt{2}$ longer than the tetragonal axes (Fig. 3-1). Also the Bi-O bonding in the Bi-O planes is more complex than found in the 2-Tl(n) materials. Further, the most studied Bi material, namely, 2-Bi(n=2), has an **incommensurate modulation**. That is, a small amount of the charge density has a periodicity along the orthorhombic **b** axis that is not an integer multiple of the **b**-axis length.

Very weak incommensurate modulations in some of the 2-Tl(n) and 1-Tl(n) materials have been detected with electron diffraction in TEM. The charge density in the Tl materials associated with the incommensurate modulation appears to be about 10× smaller than in the Bi materials. It is not yet clear if any of these distortions affect the superconductivity in any important manner or just result from the complex character of these materials. See the Notes.

Possibly the most important distortion results from deviations of the stoichiometry found in these materials. For example, in the Tl materials, some of the Ca is found in the Tl and Ba sites and vice versa. The formulae given in Table 3-1 only approximately represent the actual material, although deviations are usually slight. Certainly, deviations of the oxygen stoichiometry are known and are a subject of detailed investigations. These deviations affect the carrier concentration, which, in turn, affects T_c and other superconducting properties.

3-4 Conventional Superconductors

Most elemental superconductors have the usual structures found for the elements. These are body-centered cubic (bcc), face-centered cubic (fcc), and hexagonal close-packed (hcp) structures, and a discussion of these structures is not needed here. We mention two more complicated structures for compound, conventional superconductors.

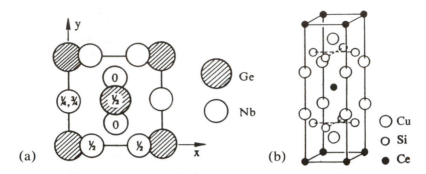

Fig. 3-9 (a) The **ab**-projection of the structure of Nb_3Ge. The atoms are all at special positions and the z-values are on the atoms, except for the Ge atom at the origin. (b) The structure of $CeCu_2Si_2$.

Nb_3Ge and related materials have the so-called **A15 structure**, shown in Fig. 3-9a. The material is cubic (space group $Pm\bar{3}n - O_h^3$) with two formula units per unit cell (Z=2). Nb_3Ge is one of a general class of A_3B materials where A can be Nb, V, Ti, Zr, or other transition metals, and B can be Sn, Al, Ga, Ge, In, or Si. T_c is sensitive to the 3/1 stoichiometry with maximum T_c values obtained for this ratio. It has been suggested that this sensitivity of T_c to stoichiometry may be associated with a sharp peak in the density of electronic states, and near the 3/1 stoichiometry the Fermi level is close to this peak.

Another interesting feature of this structure is that one can focus on "lines" of Nb atoms along all three directions. For example, the "line" of Nb atoms along the y-axis is shown at height z=0 in the figure. There has been speculation that the high values of T_c may be associated with these one-dimensional chains of atoms, which may have low-frequency phonons, which would then enhance the electron-phonon coupling parameter (Section 2-6b). However, any structure can be looked at from many different points of view. Focusing on lines of atoms is only one point of view. See the Problems.

$CeCu_2Si_2$ has the structure shown in Fig. 3-9b. It is a heavy-electron superconductor (Section 2-8c), and we mention it mostly because of its similarity to the T′ structure (Fig. 3-4a). Its space group is the same as that of the T′ structure (I4/mmm $- D_{4h}^{17}$), but the atom arrangement in the Ce plane is different. Also, note that the Cu atoms in this structure appear to have no relation to those in the T′ structure.

Problems

1. Y123 crystal structure — (a) Using any of the references given in the Notes, calculate the Cu_c-O_z, Cu_p-O_z, Cu_p-$O_p(b)$, and Cu_p-$O_p(a)$, interatomic distances. Note that for a structure, the atomic positions are quoted in terms of fractions of the unit cell dimensions. (b) What are the buckling angles for $O_p(b)$-Cu_p-$O_p(b)$ and $O_p(a)$-Cu_p-$O_p(a)$?

2. For **Nb_3Ge** (Fig. 3-9a), what are the distances between Ge and its Nb nearest neighbors? Between Nb along the "lines" of Nb atoms, and Nb and its other Nb nearest neighbors? The unit cell size is $a=5.15$Å. (From Fig. 3-9a, you should be able to deduce the coordinates of the atoms. Your result can be checked with Wyckoff or Appendix A9-5 of Burns and Glazer, Notes.)

Chapter 4

Normal State Properties

*Nature has this work to do: to shift and to change,
to remove from here and to carry there. All things are in
process of change, so that novelty should not cause fear.*

Marcus Aurelius, "To Himself"

4-1 Introduction

It is believed by many that a good understanding of the normal-state properties of high-T_C materials may significantly help lead to a deeper understanding of the superconductive properties. This belief arises because some of the metallic, normal-state properties are highly unusual. Examples of unusual properties are: The linear temperature dependence of the resistivity (i.e., $\rho \propto T$); the calculated one-electron bands for the insulator phases always, incorrectly, yield metals; effects of highly-correlated charge carriers might dominate the conduction behavior; the closeness of ordered magnetic behavior in the insulator phases. The copper oxide superconductors also have strongly anisotropic metallic properties. For example, a highly anisotropic normal-state conductivity with poor, or even nonmetallic, conductivity along the c axis, a highly anisotropic effective mass, and other anisotropic properties are expected. The questions arise as to how these properties are carried over to the superconducting state.

Fundamental questions remain about the band structure itself. In conventional one-electron band theory, it is assumed that the electrons move independently of each other. That is, it is assumed that they move in a potential that includes the *averaged* electron-electron repulsion plus the *averaged* electron-ion attraction. The Hubbard model may be used to approximate part of the electron-electron repulsion. However, for the high-T_C materials, the existence of conventional bands has been questioned. In

1987, Anderson suggested that the cuprates will not be able to be understood in terms of the types of band structures (even with correlation effects) that are appropriate to metals that are conventional superconductors. He proposed that a resonating-valence-band (RVB) type model will be required. There are many RVB variants; some tend to predict triplet-spin state, p-wave orbital state superconductivity (Section 5-2) with electron states in what would normally be the BCS superconducting gap (Section 2-5). This radical, exciting conjecture stimulated an enormous body of theoretical work (Notes).

As experimental data has accumulated, in particular, from angle-resolved photoemission measurements (PES, Sections 4-9 and 5-4a), there appears to be less of a need for exotic normal-state theories. At least for the ground-state electronic structure, band-structure calculations using, for example, the **local–density–functional approximation** (LDA), seem to be in reasonable agreement with many experiments. Thus, in this chapter, we shall discuss many normal-state properties of these high-T_c superconductors and consider them with respect to simple one-electron bands when possible. Discussions of the RVB model and even the simple Hubbard mode are left to the interested reader (Notes); these models will only be mentioned in passing when certain experimental data are discussed.

We shall discuss many of the properties of the normal states of these high-temperature superconductors. Then some of the fundamental band theory questions will be addressed.

4-2 Cu–Charge State

4-2a Charges — Atomic copper has an $[Ar]3d^{10}4s^1$ electron configuration and, in the solid state, the $4s^1$ electron is bound the weakest. In La_2CuO_4, the formal valence on the ions may be thought of as La^{3+} and O^{2-}; then we have Cu^{2+}. Thus, the Cu ion would have a $3d^9$ electron configuration. There are few compounds with Cu^{3+} ions with a $3d^8$ electron configuration, However, two closely related metals that have this electron configuration are the perovskite $LaCuO_3$ and $LaSrCuO_4$, where the latter has the same K_2NiF_4 structure as does La_2CuO_4. The closed shell $3d^{10}$ electron configuration is found in compounds like Cu_2O (Cuprite).

La_2CuO_4 is an insulator, but when sufficiently doped with Sr, it becomes a metal, Sr acting as an accepter, since Sr^{2+} ions give up one less electron than La^{3+} ions. Keeping La and O with formal ionic charges of 3+ and 2-, then the formal Cu ionic charge in $(La_{2-x}Sr_x)CuO_4$ is

Cu^{2+x}. Theoretical calculations could help to tell where the charge actually resides. Still allowing only the formal charge of Cu to vary, descriptions for the doped material have also been written as $(1-x)Cu^{2+}$ and xCu^{3+}. However, it is not clear if such a description helps the understanding.

The Y123 material also has charge states that vary with doping. The insulator $YBa_2Cu_3O_6$ has two four-coordinated Cu^{2+} ions in the plane, and a single two-coordinated Cu^{3+} ion on the sticks (Section 3-2f). However, the superconductor $YBa_2Cu_3O_7$ has three $Cu^{2.33+}$ ions if the charge is divided equally among the Cu ions and still assuming that the excess charge resides only on the Cu ions.

Applying the same type of charge counting to the other superconducting cuprates (Table 3-1) results in similar formal charges higher than two on the Cu ion in many cases. Thus, these cuprate metals are not traditional metals from a formal ionic point of view.

4-2b Molecular Orbitals — Appreciating the formal ion charges is helpful, but the structural backbone of the superconducting cuprates is the strongly covalent Cu-O planes. The short Cu-O distances in these planes indicate covalent and not ionic bonding. Consider the bonding of a full Cu-O octahedron (CuO_6); that is, from the 3d orbitals on the Cu ion bonding with the 2p orbitals on the surrounding O ions. Actually, the in-plane $Cu-O_p$ distance is about 1.9Å compared to the out-of-plane $Cu-O_z$ distance of about 2.4Å (Section 3-2e). Thus, the $Cu-O_z$ covalent bond is largely nonbonding compared to the in-plane $Cu-O_p$ bonds.

For a Cu-O octahedron in the La(n=1) structure, there are 17 orbitals. Five are from the 3d orbitals on Cu, which are $d(x^2 - y^2)$, $d(z^2)$, and the three $d(xy)$ type. The Cu atoms have $2O_z$ atom neighbors along the **c** axis, above and below, and in the unit cell only $2O_p$; the other two nearest neighbor planar oxygen atoms must be considered in the next unit cells (Fig. 3-1). These four O atoms each have three p orbitals, which contributes 12 orbitals for a total of 17 orbitals.

The secular determinant for these 17 orbitals is not difficult because it block diagonalizes into 3×3 determinants at most. However, we will focus on the in-plane covalent bonding, and take a more intuitive approach, as is often done in molecular-orbital theory (Notes). To do this, consider the two planar O atoms with p orbitals that are directed toward the central Cu atom. On the central Cu atom, we only use the $d(x^2 - y^2)$ orbital, since it is correctly oriented for σ bonding with its neighboring oxygens. Thus, only these three orbitals are used and, to a first approximation, the other 14 orbitals can be taken as nonbonding relative to these orbitals. The principle

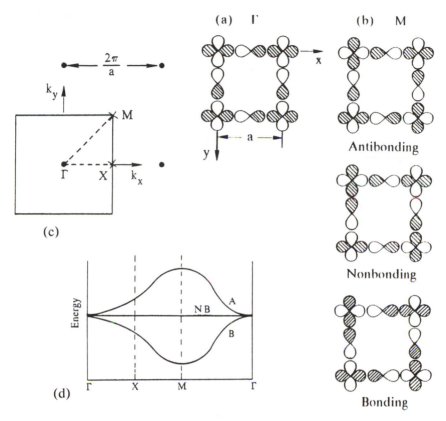

Fig. 4-1 (a) A Cu-O plane, viewed from above, shows σ bonding between the Cu $d(x^2 - y^2)$ and oxygen p_x and p_y orbitals. The repeat distance $=a$; therefore, the bonding refers to $k=0$ or the Γ point in the Brillouin zone. (b) is the same except corresponding to the M point in the Brillouin zone. (c) The reciprocal lattice (solid dots) of a square lattice with the first Brillouin zone outlined by solid lines. The special points Γ, X, and M are indicated, as is the irreducible wedge with dashed lines. (d) A sketch of E vs. **k** for the bonding (B), antibonding (A), and nonbonding (NB) orbitals.

being applied here is that maximum interaction occurs for orbitals that have maximum overlap.

Figure 4-1a shows a Cu-O plane with $d(x^2 - y^2)$ orbitals attached to Cu atoms, and with p_x and p_y orbitals attached to the O atoms. The orbitals on Cu are all the same, and those of the O atoms are the same in the **a** and **b** directions, so that the repeat distance is the unit cell size ($=a$). Thus, this molecular orbital diagram corresponds to the $k=0$ or Γ point in the Brillouin zone. Bonding between charge-lobes of the orbitals occurs when the orbitals have the same phase (shaded-shaded or white-white). Anitbonding occurs between orbitals with opposite phase (a shaded and a white orbital). For the molecular orbital in Fig. 4-1a, there are as many bonding as anitbonding overlaps. Since the bonding and antibonding energies are approximately the same but opposite in sign, this molecular orbital is nonbonding.

The orbital diagram in Fig. 4-1b has a repeat distance $=2a$ in both the x and y directions; this is the M point in the Brillouin zone. The Cu-O plane has a square lattice and the first Brillouin zone is shown in Fig. 4-1c, where the Γ, X, and M special points are indicated.

Note, at the M point, three different types of orbitals can be drawn (Fig. 4-1b). The bonding orbital (B) has the maximum amount of overlap and hence the lowest energy. The antibonding orbital (A) has the minimum amount of overlap and hence the highest energy. Last, the nonbonding orbital (NB) has as many bonding as antibonding overlapping orbitals; hence, the overlap causes little energy change from the atomic case. The other 14 orbitals we take as nonbonding, so they have energy levels that are essentially independent of k. Of course, π bonding could be considered, but it is weaker than σ bonding and will be ignored here, and bonding to the distant O_z atoms is also ignored to a first approximation.

Figure 4-1d is a sketch of E vs. k around the irreducible wedge of the Brillouin zone. The bonding, nonbonding, and antibonding states are indicated. For clarity, we have not sketched the other 14 bands, which we take as essentially nonbonding. However, if all of the wavy lines that belong in Fig. 4-1d, from the 17 bands, are drawn, these diagrams are sometimes fondly referred to as "spaghetti diagrams."

How many electrons do we have to put into these 17 orbital bands? Well, Cu contributes 11 electrons from its $3d^{10}4s^1$ configuration. Each O contributes four electrons from its $2p^4$ configuration, and there are four O atoms in the unit cell of our octahedral complex; thus, these oxygen atoms contribute 16 electrons. Last, we take La^{3+}, so the two La ions in the unit cell contribute six electrons. Thus, in the Cu-O molecular orbitals, there are $11+16+6=33$ electrons.

Summarizing, there are 17 orbitals, each of which can accommodate two electrons (spin up and down), and 33 electrons to be included. Thus, all of the orbitals will be filled except for the highest-energy, antibonding one labeled A in Fig. 4-1d. This orbital is half filled. The energy of this orbital varies with k, as in Fig. 4-1d. So, the orbital will have occupied k states from approximately the Γ to X points, and unoccupied states from approximately the X to M points. The occupied states lie below the Fermi energy E_F and the unoccupied states lie above E_F, and this is again discussed in Section 4-7e.

In the atomic state, the ionization energy of a p electron from the $2p^4$ O orbitals is larger than that of a 3d Cu orbital. Thus, the p orbitals of O are lower in energy (larger binding energy) than the d orbitals of Cu. Therefore, the bonding orbital B is mostly O p-like, and the antibonding orbital A is mostly Cu $d(x^2 - y^2)$-like. Thus, we have arrived at the important point of this discussion. Near E_F, the top of the conduction band, the orbital character is mostly Cu $d(x^2 - y^2)$-like.

4 – 3 Resistance

4-3a Conventional Resistivity Behavior — Before discussing the anisotropic resistivity found in the normal state of high-T_c crystals, we discuss what is found in simple, classical metals. Resistivity vs. reduced temperature, for some simple metals, is shown in Fig. 4-2a. The reduced temperature is T divided by the Debye temperature (Θ_D). As can be seen for these simple metals, a universal curve results and the resistivity can be divided into a low and high temperature part.

For the high temperature part,

$$\rho = A + BT \qquad \text{where } T \gtrsim 0.2 \, \Theta_D \qquad (4 - 3a)$$

In this classical-temperature region, the details of quantization of the lattice vibrations, which cause resistance, are not important. Rather, the electron scattering is proportional to the square of the amplitude of the atomic vibrations about their equilibrium positions. Calculations of the resistivity, and other transport properties of metals in the high-(Eq. 4-3a) and low-temperature regime, are not difficult, but they are long and cumbersome. We continue to discuss the resistivity qualitatively and refer the reader to the Notes for references to quantitative discussions.

A detailed understanding of the low-temperature regime is more complicated than the high-temperature regime. This is because at low temperatures, the electron scattering cannot be analyzed in terms of classical departures of the atoms from their equilibrium positions. Rather, the electron scattering from quantized lattice vibration (phonons) must be considered. The phonons have energy $\hbar\omega_q$ and crystal momentum $\hbar q$. The

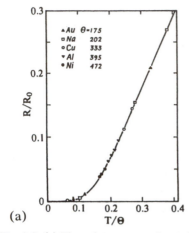

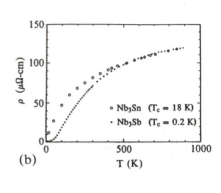

(a)

(b)

Fig. 4-2 (a) The points are experimental resistivity data plotted vs. reduced temperature, for the various metals as indicated. The solid line is a result from the Grüneisen-Bloch formula. (b) The temperature dependence of the resistivities of two conventional superconductors as labeled. For more details, see the text and the article by Z. Fisk and G. W. Webb in the book by D. H. Douglass, Ed. (Bib.).

phonon contribution to the electrical resistivity arises from the scattering of electrons by phonons of the type,

$$\mathbf{k_i} + \mathbf{q} \overset{\leftarrow}{\underset{\rightarrow}{}} \mathbf{k_f} \qquad\qquad (4-3b)$$

where $\mathbf{k_i}$ and $\mathbf{k_f}$ are the initial and final electron $\mathbf{k}$ values. At low temperatures, the combination of both small $\mathbf{q}$ and conservation of crystal momentum (Eq. 4-3b) strongly reduces the possible channels for electron scattering, which greatly reduces the resistance. At low temperatures,

$$\rho \propto T^5 \qquad \text{where } T \ll 0.2\,\Theta_D \qquad\qquad (4-3c)$$

is obtained, which can be qualitatively seen in Fig. 4-2a.

The **Grüneisen–Bloch formula** extrapolates between the low and high temperature limits, and is the solid line in Fig. 4-2a. Note: Eq. 4-3c suggests that the resistivity goes to zero at absolute zero because, at this temperature, no phonon states are populated. This result is appropriate, since the calculations only consider the contribution of phonons to the resistivity. That is, the Bloch theorem, encountered in the usual band theory discussions, leads to perfect (i.e., zero) electrical conductivity, since there is no scattering of the electrons from one quantum state to another (of the type indicated in Eq. 4-3b) because the theorem applies to electrons propagating in a perfect, periodic structure. In order to obtain a non zero resistance, mechanisms must exist that scatter electrons between their (sharp) quantum states (i.e., mechanisms that give rise to scattering of the type in Eq. 4-3b). The electron-phonon interaction is just one mechanism that gives rise to a finite resistance for T>0. The interaction of electrons with other deviations from perfect translational symmetry also gives rise to electron scattering and a finite resistance. Typically, these other deviations may include other electrons, point or line defects, a small percentage of magnetic impurities, which, due to the long-range interaction between their localized magnetic moment and that of the electron, are particularly effective in scattering electrons. Most of these other deviations from perfect translational symmetry are static and hence yield a temperature-independent resistivity (equal to C). Hence, at low temperatures, a resistivity of the form,

$$\rho \approx C + DT^5 \qquad (4-3d)$$

should be expected, where the second term comes from Eq. 4-3c.

The temperature dependence of the resistivities of two conventional superconductors with the A15 structure (Section 3-4) are shown in Fig. 4-2b. At low temperatures, $\rho(T)$ for Nb_3Sb is similar to that shown in Fig. 4-2a for elemental metals. At very high temperatures, both A15 compounds have nearly identical resistivities. However, at lower temperatures, $\rho(T)$ of Nb_3Sn is considerably higher than Nb_3Sb, which is ascribed to the stronger electron-phonon parameter in the former. Thus, we see how λ_{ep} (Section 2-6) not only can effect $\rho(T)$ in the normal state, but also strongly increase T_c (Eq. 2-6b).

4-3b   Resistivity of High-T_c Materials — With this overview of classical-metallic resistivity, we now turn to results in high-T_c crystals. Different behavior is found.

The resistivity of single-crystal Y123 perpendicular to the c axis, that is, in the **ab** plane, is shown in Fig 4-3a. Actually, ρ_a and ρ_b are measured separately. If the crystal is multidomain, then one measures ρ_{ab}. Over the measured temperature range, $\rho \approx E + FT$, which is unusual in a metal. More usually, the low-temperature resistivity is closer to $\rho = A + BT^5$, and at high temperatures, the behavior is linear with temperature as discussed in Section 4-3a. More usually, ρ_{ab} is measured and in fully oxygenated Y123, and this approximately linear behavior is found.

For the c-axis resistivity (ρ_c), results are somewhat controversial. The fact that both ρ_{ab} and ρ_c are linear with temperature suggests that Y123 may be regarded as a highly anisotropic three-dimensional metal, rather

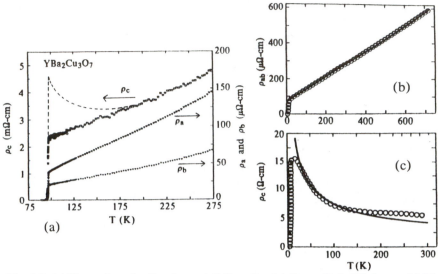

Fig. 4-3 (a) The a-, b- and c-direction resistivities. For details, see T. A. Friedmann, M. W. Rabin, J. Giapintzakis, J. P. Rice, and D. M. Ginsberg, Phys. Rev. B **42**, 6217 (1990). The dashed curve is a schematic result for ρ_c; see the text for a discussion. (b) and (c) ρ_{ab} and ρ_c for a 2-Bi(n=1) superconductor with $T_c \approx 7$ K. For details, see S. Marten, A. T. Fiory, R. M. Fleming, L. F. Schneemeyer, and J. V. Waszczak, Phys. Rev. B **41**, 846 (1990).

than a two-dimensional metal. Other published measurements along the c axis show upward curvatures above T_c, as sketched by the dashed line (Fig. 4-3a), which suggests some sort of localization. It has been argued that the linear behavior of ρ_c occurs in the best crystals (and contacts). However, the meaning of "best" is not clear, since many fully oxygenated Y123 crystals with very sharp ρ vs. T at T_c still show an upward curvature similar to that sketched (Fig. 4-3a). It is possible that in Y123 along the c axis, the material is close to a metal-insulator transition, and small changes in composition or defects determine which takes place. However, an understanding between Y123 single crystals that along the c axis show linear ρ_c behavior and those that show localization behavior is not well understood (but see Notes for more recent work).

The results in Fig. 4-3a yield an approximate temperature-independent ratio, $\rho_a/\rho_b = 2.2$, so the chains make a large contribution to the **ab-** plane *conductivity*. Naively separating the chain and plane conductive as:

$$\sigma_b = \sigma_{chain} + \sigma_{plane}$$
$$\sigma_a = \sigma_{plane} \tag{4-3e}$$

Then the room temperature conductors are:

$$\sigma_{chain} = 0.0084(\mu\omega - cm)^{-1}$$
$$\sigma_{plane} = 0.0054(\mu\omega - cm)^{-1}$$
$$\frac{\sigma_{chain}}{\sigma_{chain} + \sigma_{plane}} = 0.60 \tag{4-3f}$$

Thus, at room temperature, in Y123, more that 50% of the dc conductivity may be due to the chains. The implication of this result for the superconductivity in this material is, however, not clear.

The results in Fig. 4-3a are, of course, from twin-free crystals. To compare to **ab**-plane measurements of twinned crystals, probably the geometric mean is appropriate to take, $\rho_{ab} = (\rho_a\rho_b)^{1/2}$ whi $\approx 110 \ \mu\Omega$-cm at room temperature. This results are about 1.5 to 2 lower than reported in the "best" twinned crystal. This may indicate the significance of twin boundaries in the **ab**-plane resistivities. These results for ρ_c/ρ_b (Fig. 4-3a) vary from ~75 to ~120 at 275 K and 100 K, respectively. Thus, Y123-O_7 has a highly anisotropic resistivity.

Figure 4-3b shows $\rho_{ab}(T)$ for 2-Bi(n=1). For this particular crystal, $T_c \approx 7$ K and, as can be seen, $\rho_{ab} = A + BT$ over an extremely wide temperature range (about 7 to 700 K). For this same crystal, ρ_c is shown in Fig. 4-3c; a power law behavior gives a reasonable fit to the data, $\rho_c \propto T^{-\alpha}$, where $\alpha \approx 0.61$. In fact, a similar power law dependence is always found for ρ_c in 2-Bi(n=1 and 2) crystals. Although the physics underlying a power law behavior is not clear, $\rho_c(T)$ does not have a thermally activated form as would be expected for semiconductor-like behavior. The results in Fig. 4-3c resemble the dashed ("localization") result in Fig. 4-3a.

For 2-Bi(n) crystals, ρ_c/ρ_{ab} is always much larger than values found in Y123, and temperature-dependent ratios between 500 and about 10^6 have been reported. In 2-Bi(n) crystals, there are four isolation planes (2Ba-O and 2Bi-O planes) between each group of n Cu-O planes (Chapter 3). However, in Y123, between the 2 Cu-O planes there are only three isolation planes (2Ba-O and a Cu-O plane that forms the chain — Chapter 3). This difference in the number and type of isolation planes may account for part of the different ρ_c/ρ_{ab} ratios. However, since the chains in Y123 appear to be good conductors (Eq. 4-3f), this thought may be meaningless. However, comparison of 1-Tl(n) and 2-Tl(n) crystals may prove interesting.

The resistivity and anisotropy in La(n=1) are similar to those in Y123. In $(La_{2-x}Sr_y)CuO_4$, the doping is relatively easily controlled by adding Sr atoms, which increases the number of holes. For $x \lesssim 0.06$, the material is an insulator. For $x \gtrsim 0.06$, it is a metal and a superconductor. In the metallic region, up to $x \approx 0.15$, T_c increases as the hole concentration increases (to be discussed later). For higher x values, T_c decreases until x $\gtrsim 0.26$ and the material is no longer a superconductor. However, for these large x values, the material remains a metal. In fact, the conductivity is even higher than for x values that show superconductivity; this effect is similar to that shown in Fig. 2-1b. Actually, we note that the linear ρ_{ab} vs. T is hardly exact. In fact, deviations from straight line behavior can be seen in Fig. 4-3a for both ρ_a and ρ_b above 200 K. Further, to indicate the variety of unusual effects that can be found in other layered superconductors, we note that in the layered-organic superconductors (Section 2-8d), $\rho_{ab} = G + HT^2$ has been reported in the normal state of some of these superconductors (Notes).

Return now to the in-plane resistivity and the form, $\rho_{ab} \approx A + BT$, found for many high-T_c crystals. It is interesting to note that most of the high-T_c crystals and oriented films yield similar B values, B ~ 1 $\mu\Omega$-cm/K. This suggests a common scattering mechanism for carrier transport in the Cu-O planes. This common B value gives hope for a universal understanding of these materials. However, the understanding must be different from that found in normal metals of the type discussed in Section 4-3a. For ordinary metals, we only expect linear ρ_{ab} behavior for $T \gtrsim 0.2\Theta_D$ (Eq. 4-3a). The **mean free path**, average distance between scattering of the carriers in the **ab** plane in Y123 at about 100 K, is estimated to be about 100Å to 200Å. As we will see later (Section 5-7d), this puts the superconductors in the "clean limit."

One explanation of the linear-T behavior has been found by considering the resistivity to be two-dimensional (i.e., **ab** plane) and due to two carries (could be electors and holes, for example) that interact with each other. This model is consistent with highly anisotropic effective masses that are expected for high-T_c materials because of their structure. An interesting feature of this model is that it gives a quantitative explanation of the absence of intrinsic EPR (electron-paramagnetic resonance) in high-T_c crystals. Since high-T_c crystal should contain Cu^{2+}, strong **EPR signals** are expected, but have not been observed. This two-band, two-dimensional model predicts a spin-flip relaxation term due to carrier interactions that is orders of magnitude shorter than the EPR-signal period, making the signal essentially impossible to observe because of broadening (Notes).

To summarize, many high-T_c superconductors show an approximate linear temperature dependence for ρ_{ab} ($\approx A+BT$) with similar B values, perhaps indicating a common origin for the temperature term, but see Section 4-7e. Thus, the high-T_c cuprates have a normal-state ρ_{ab} behavior that is different from that of the common metals (Fig. 4-2). More variation is found for ρ_c values. Upward curvature (localization) behavior is found for some Y123 crystals and all 2-Bi($n=2$ and 3) crystals. Some theoretical justifications have been given.

4 - 4 Hall Effect

It is more difficult to obtain Hall results than resistivity results; hence, there are, unfortunately, fewer measurements. It would be useful to be able to consider only single-crystal results, but there are not enough of these measurements. Fortunately, dense polycrystal samples usually yield Hall effects with the same signs as those found in single crystals when the magnetic field is along the **c** axis.

Hall measurements yield the **Hall coefficient** R_H, which is determined when the magnetic field and current are perpendicular to each other, with the Hall voltage measured in the third orthogonal direction. For carriers in a single parabolic band, elementary calculations show that R_H is related to the carrier density N/V as

$$R_H = \frac{-1}{e(N/V)} \quad \text{(SI)}$$
$$= \frac{-1}{ec(N/V)} \quad \text{(cgs)} \qquad (4-4a)$$

where e is the proton charge and c the velocity of light. For simplification, instead of stating R_H, it can be more useful and intuitive to quote the **Hall density** or **Hall number** N/V obtained by inverting Eq. 4-4a. This is just the carrier density, which for metals such as Cu is $\sim 5 \times 10^{22}/cm^3$. This carrier density may be quoted in terms of the number of carriers per planar Cu atom. However, this should not imply any significance to the physical interpretation; it is only that quoting N/V usually provides the nonspecialist with a value that is easier to understand and remember. The **Hall mobility** is defined as $\mu_H = (R_H)(\rho)$, where ρ is the resistivity.

If both electrons and holes are present in parabolic bands with a density of n and p, respectively, and writing (in SI units) $R_n = -1/ne$ and $R_p = 1/pe$, then the **two-band Hall coefficient** is

$$R_H = \frac{1}{e} \frac{p\mu_p{}^2 - n\mu_n{}^2}{(p\mu_p + n\mu_m)^2} \qquad (4 - 4b)$$

The measured R_H lies between R_n and R_p. In this case, $(eR_H)^{-1}$, the Hall density, is only the upper limit of the electron or hole density. If the density of electrons and holes as well as their mobilities are temperature-independent, then the two-band R_H will be temperature-independent.

If the band structure is complicated, then even for the single carrier case (Eq. 4-4a) the Hall coefficient cannot be simply written in terms of the carrier concentration. Rather, it involves a complicated integral over the Fermi surface. Thus, R_H may not be easy to interpret, and this seems to be the case at least in Y123. Nevertheless, for a given material, the variation of the Hall coefficient with doping may give important information about the effectiveness of the doping.

La(n=1) — R_H vs. T for $(La_{2-x}Sr_x)CuO_4$ is shown in Fig. 4-4a. The very strong temperature dependence observed for x=0 is largely absent for this material, indicating that a Hall number can be safely obtained. The units in Fig. 4-4a are $cm^3/Coulomb$; to obtain the Hall number, the inverse of R_H must be divided by the proton charge = 1.60×10^{-19} C. The sign of the Hall measurements indicates that *the carriers are holes,* as should be expected when La^{3+} is replaced by Sr^{2+}. Approximately one hole is added for each Sr atom, at least up to $x \approx 0.1$. For x=0.1, the Hall number, via Eq. 4-4a, yields ≈ 0.1 holes per planar Cu atom (a value consistent with the chemical (redox titration) method used to determine the hole concentration). In this body-centered tetragonal structure (Fig. 3-1a), there are two Cu atoms in a unit cell of size $\approx 3.79 \times 3.79 \times 13.29 \text{Å}^3$, so, via Eq. 4-4a, the carrier concentration is $\approx 1 \times 10^{21}/cm^3$.

For $(La_{2-x}Sr_x)CuO_4$, R_H is relatively independent of temperature up to $x \approx 0.1$ with $(N/V) \propto x$, so $R_H \propto x^{-1}$, as can be seen in Fig. 4-4b. At x=0.15, which yields the highest T_c value, R_H is about two times smaller than expected from the x value (Fig. 4-4). For larger x values, R_H becomes even smaller and changes sign for $x \gtrsim 0.3$, which approximately coincides with the disappearance of superconductivity. T_c vs. x is shown in Fig. 4-3a, which shows the typical bell-shaped curve found for most high-T_c materials (which looks flattened because log T is plotted). These typical results show that there is an optimimal doping to obtain the maximum T_c.

Electron-doped material — $(Nd_{2-x}Ce_x)CuO_4$ is a superconductor with the T' structure (Fig.3-4a). The sign of the Hall measurements indicates that *the carriers are electrons,* as would be expected when La^{3+} is replaced by Ce^{4+}. (This is for H parallel to c so that the carriers are circu-

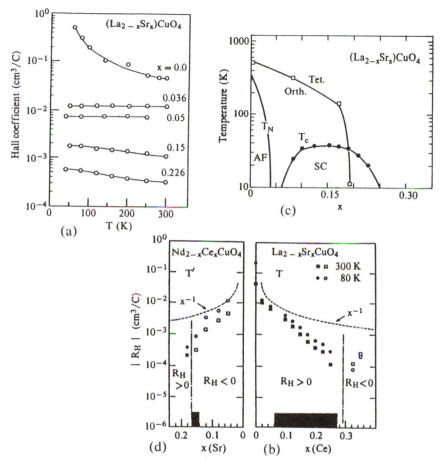

Fig. 4-4 (a) R_H vs. T for La(n=1) with different Sr concentrations. (b) $|R_H|$ vs. Sr content in La(n=1). See H. Takagi, T. Ido, S. Ishibashi, M. Uota, S. Uchida, and Y. Tokura, Phys. Rev. B **40**, 2254 (1990). (c) Temperature vs. composition showing the tetragonal and orthorhombic structure regions, the antiferromagnetic (AF) region, the Néel temperature T_N, and superconductor (SC) regions for $(La_{2-x}Sr_x)CuO_4$. See J. B. Torrance, A. Bezinge, A. I. Nazzal, T. C. Huang, S. S. Parkin, D. T. Keane, S. L. LaPlaca, P. A. Horn, and G. A. Held, Phys. Rev. B. **40**, 8872 (1989). (d) This figure is similar to that in (b), but here the material is Nd(n=1). For details, see A. Kitazawa in Bednorz and Müller, Eds. (Bib.).

lating in the **ab** plane). Figure 4-4d shows the experimental results for this crystal, grown in a reducing atmosphere. The Hall coefficient is negative and is approximately given by x^{-1} up to x=0.1. Thus, up to x=0.1 and using Eq. 4-4a, $N/V \propto x$. For larger x values, R_H rapidly becomes smaller and, for $x \gtrsim 0.16$, it changes sign (to hole carriers) and is very small. This material is superconducting in only a narrow range of x, as can be seen in Fig. 4-4d.

The symmetry between these results in the hole and electron doped superconductors, which can be seen in Figs. 4-4b and 4-4d, is striking. This symmetry appears to imply that the pairing mechanism in these materials is independent of the sign of the charge carrier. Further, it appears that superconductivity occurs in a region of doping where the sign of the carriers is about to change from what is expected to the opposite sign.

Y123 — $YBa_2Cu_3O_{7-\delta}$ is a more extensively studied material and the anisotropy in R_H has been measured. However, the results are more complicated in that temperature-dependent R_H values are generally found. Thus, it is more difficult to interpret the results.

First, consider **H** parallel to **c**, so the current and voltage is from carriers in the **ab** plane; the Hall effect is positive, so the carriers are holes as found for La(n=1). The temperature dependence that is usually found is $(R_H)^{-1} \propto T$. Y123 has Cu atoms in the chains as well as in the planes (Fig. 3-3), and it is not clear if the holes are located on the Cu or O atoms in the Cu-O planes, or on the Cu or O atoms in the chains. It has been suggested that holes on the chains may be immobile. Other considerations in Y123 are the easily variable O concentration and, as always for these superconductors, the possibility of complicated band effects. At least, the Hall number generally decreases with increasing δ, which is consistent with the decreasing negative charge from fewer O^{2-} ions. Using $\rho \propto T$ and $(R_H)^{-1} \propto T$, a Hall mobility $\mu_H \propto T^{-2}$, which at $T \approx 100$ K is $\mu_H \sim 34$ cm^2/ sec -V, a small mobility.

Hall measurements have been made on compensation-doped Nd123; that is, the Y^{3+} is replaced by Nd^{3+} plus a few percent of Ca^{2+} atoms. In this Nd123 (still with $T_c \approx 90$ K), R_H shows a much smaller temperature dependence that tends to flatten near 100 K (Fig. 4-5a). In this temperature region the hole density using Eq. 4-4a and that obtained from the wet chemical redox titration method are in agreement with each other. Then the Hall number is about 0.2 holes per planar Cu atom. Also, ~0.2 holes per planar Cu atom is in reasonable agreement with what is obtained from Y123 at T~100 K. In the primitive unit cell of Nd123, there are two Cu atoms in the Cu-O plane with a cell size of $3.82 \times 3.89 \times 11.68$Å^3, so the carrier concentration is $\approx 2.3 \times 10^{21}$/cm^3. These carrier densities are low compared to conventional metals, being an order of magnitude smaller than the lowest values found in elemental metals. T_c vs. delta is shown in Fig. 4-5b (and is

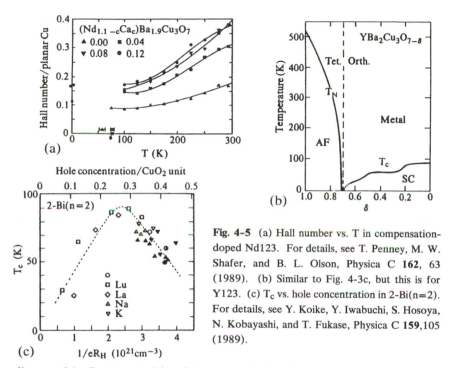

Fig. 4-5 (a) Hall number vs. T in compensation-doped Nd123. For details, see T. Penney, M. W. Shafer, and B. L. Olson, Physica C **162**, 63 (1989). (b) Similar to Fig. 4-3c, but this is for Y123. (c) T_c vs. hole concentration in 2-Bi(n=2). For details, see Y. Koike, Y. Iwabuchi, S. Hosoya, N. Kobayashi, and T. Fukase, Physica C **159**,105 (1989).

discussed in Section 5-6h). Of course, δ describes the oxygen content in Y123 and is possibly not simply related to the number of holes. However, it appears that $T_c = 92$ K is obtained for ~0.2 holes per plane-Cu atom (Fig. 4-5a), and T_c decreases as the doping is changed from this value.

For Y123 crystals, when **H** is in the **ab** plane, the Hall coefficient has the opposite sign (electron-like carriers) from when **H** is parallel to **c**, as described before. Also, for **H** in the **ab** plane, N/V (Eq. 4-2a) increases upon cooling and is about 2 to 5 larger than the 0.2 per planar Cu atom. These results in Y123 are unusual and show the highly anisotropic nature of the electronic conduction.

There have been a number of first-principle-type calculations of the Hall coefficient based on different types of assumptions about the bands. Some of these calculations in Y123 indeed obtain positive and negative Hall coefficients for **H** parallel and perpendicular to the **c** axis, respectively, in agreement with experiment. The same change in sign is also calculated in La(n=1). However, most of these calculations find temperature-independent R_H values. In order to obtain R_H values with T^{-1} temperature dependence, a very narrow band tends to be required.

Other crystals — The Hall effect has been reported in 2-Bi($n=2$) and 2-Tl($n=2$ and 3) for **H** parallel to **c**. The magnitudes of R_H are similar to those found in Y123, but the temperature dependence is less and the signs are all positive (hole-like carriers). Near 100 K, the Hall numbers are about 0.35, 0.10, and 0.15 holes per Cu atom, respectively.

Thus, the temperature dependence for R_H found in Y123 appears to be more of an exception rather than the rule; perhaps it is associated with the chains that are present in Y123. In $Bi_2Sr_2CaCu_2O_8$, 2-Bi($n=2$), it is possible to replace Ca^{2+} in between the Cu-O planes with Y^{3+}, Lu^{3+}, and Na^+. Also, the Sr^{2+} can be replaced by La^{3+} and K^+. With all of these atom replacements, the number of holes varies and Fig. 4-5c shows T_c vs. Hall number. The typical bell-shaped curve results with T_c peaking at 0.2 to 0.3 holes per planar-Cu atom. Another, less controllable way to vary the doping is to force extra oxygen into the structure at interstitial sites. This is usually done by annealing in high-pressure O_2 gas at elevated temperatures. For example, in 2-Tl(2=1), T_c can be varied from 0 (metallic) to 85 K; the more interstitial the oxygen, the lower T_c. The excess oxygen appears to reside between the two Tl-O planes. See the Notes for Chapter 3.

4 - 5 Magnetism

4-5a Insulator Phase — In the insulator phases of all of the high-T_c materials, the Cu ion magnetic moments order antiferromagnetically at relatively high temperatures. For undoped La($n=1$), the Néel temperature is $T_N \approx 340$ K. Figure 4-6 shows the ordering of the Cu magnetic moment; within a Cu-O plane, the nearest-neighbor Cu moments are aligned in opposite directions.

Y123 is insulating for $\delta \approx 1$ and $T_N \approx 500$ K. The antiferromagnetic ordering of the magnetic moments on the planar Cu atoms is essentially the same as in La($n = 1$), shown in Fig. 4-5; adjacent spins in the plane are antiparallel. For $\delta = 1$, the Cu atoms in the chains (Cu^{3+}) are nonmagnetic.

Figures 4-4c and 4-5b show the dependence of T_N with doping in La($n=1$) and Y123, respectively. The Néel temperatures decrease rapidly as the metallic phases are approached. These figures also show how T_c is affected by doping, which will be discussed again in Chapter 5. For La($n=2$), a spin-glass phase has been reported between the phases labeled AF and SC in Fig. 4-4c.

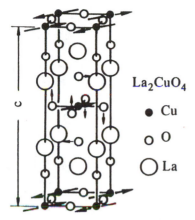

La_2CuO_4

● Cu

○ O

◯ La

Fig. 4-6 A unit cell of La_2CuO_4. The larger black arrows on the Cu atoms denote the spin directions below T_N. The smaller arrows attached to the O atoms surrounding the central Cu atom indicate the tilts that cause the tetragonal material to become orthorhombic at low temperatures. Also see Fig. 3-7.

Many of the other high-T_c materials listed in Table 3-1 also can be made insulating, for example, by partially replacing the Ca^{2+} between the Cu-O planes with Y^{3+}, or vice versa, depending if you start with 2-Tl(n) or Y123. For example, $Bi_2Ba_2(Ca_{1-x}Y_x)Cu_2O_8$ is insulating for x $\gtrsim 0.5$. In the insulating phases, these materials also become antiferromagnetic.

4-5b Superconducting Phase — In some rather special conventional superconductors, antiferromagnetic order is observed below T_c without drastically affecting the superconductivity (Section 2-8a). In the high-T_c materials, the coexistence of antiferromagnetism at temperatures well below T_c is not unusual; for example, in Y123 the Y atoms can be replaced by rare earth atoms (with local f-shell magnetic moments) with essentially no effect on T_c; in fact, T_c increases slightly, perhaps up to 95 K, which is probably associated with ionic-size effects. Then, at much lower temperatures, the rare-earth atoms can magnetically order with no effect on the superconductivity. For example, in $GdBa_2Cu_3O_{7-\delta}$, the Gd spins order antiferromagnetically at 2.2 K independent of whether the material is a metal or an insulator. Apparently, the superconducting wavefunction does not overlap the Y-sites because if it did, the large fluctuating Gd magnetic moment would be expected to strongly decrease the electron-pairing energy. This would lead to Gd123 having a (much) lower T_c than Y123, the situation found in most conventional superconductors (Section 2-8a). For some of the high-T_c crystals, where T_c and T_N are closer to each other, the situation may be more complicated. For example, consider the electron-doped superconductors with the T' structure (Section 3-4a). Pure (non-superconducting) Sm_2CuO_4 has $T_N = 5.95$ K. For superconducting $Sm_{1.85}Ce_{0.15}CuO_4$ ($T_c \approx 11.5$ K), the Sm ions order antiferromagnetically, with $T_N \approx 4.9$ K. At this temperature, there is a change in slope of H_{c2}, reminiscent of what is sometimes observed in conventional superconductors (Section 2-8a).

For conventional superconductors, magnetic impurities, or external magnetic fields tend to destroy superconductivity. In this spirit, it also appears that a magnetic moment on the Cu atom site and superconductivity are mutually exclusive (Fig. 4-5). On the other hand, it can be argued that the close proximity of antiferromagnetic and superconductive states may indicate a common origin, or at least a closely related origin, and this is discussed in Section 4-7. This latter thought has led to the idea that the electron-electron attraction required to form Cooper pairs might arise not from phonons but from some kind of antiferromagnetic magnons (i.e., magnon-mediated pairing), and these ideas are being studied.

The $BaBiO_3$-type superconductors are not antiferromagnetic in their insulating phases, and superconductivity is not observed above 30 K. Also, the crystal structure, the magnitude of the superconducting energy gap, the isotope effect, and coherence length appear to be different from the planar cuprate materials. Thus, the fundamental pairing mechanism may be different. For these reasons, we continue to concentrate on high-T_c materials (with Cu-O planes) and only consider the $BaBiO_3$-type materials in passing.

4-6 Structural Phase Transitions

Besides an antiferromagnetic phase transition in $(La_{2-x}Sr_x)CuO_4$, there is also a structural phase transition from a high-temperature tetragonal structure to a lower-temperature orthorhombic structure. The x dependence of the structural-transition temperature is shown in Fig. 4-4c; however, exactly where the structural phase transition crosses the superconducting phase is controversial. The atomic movements causing the low-temperature structure are shown in Fig. 4-5 for just the central Cu atoms and its six oxygen neighbors. The motion can be described as the tilting (rotating) about the [110] direction of the octahedra composed of the Cu-O plane atoms and the two O_z atoms. Due to the fact that the Cu-O planes share O atoms, all of the oxygen atoms in the plane tilt either up or down. This motion is related to a soft Brillouin zone boundary phonon that drives the structural phase transition. A more complete discussion is in Section 3-3.

$(La_{2-x}Ba_x)CuO_4$ has a phase diagram similar to that shown in Fig. 4-4c, but there is another structural phase transition at temperatures in the neighborhood of T_c to a low temperature tetragonal (LTT) unit cell. In this

LTT phase, the octahedra are alternately tilted about either the **a** or **b** axes. At present, the relationship of these structural phase transitions and superconductivity is not clear, however; see Section 3-3.

Y123 is not known to have any structural phase transitions, at least in the conventional sense of one occurring at constant composition. However, as indicated in Fig. 4-5b, when the material is oxygen deficient, it is tetragonal and an insulator. When the material has more oxygen, it is an orthorhombic metal and a superconductor. However, there are occasional diffraction reports of large thermal factors of some of the oxygen atoms, namely, O_c(b) and sometimes O_z (Fig. 3-3). Some of these data are interpreted as static or dynamic ordering of these atoms with small displacements (~ 0.2Å) from their normally reported sites. Future studies in this area should prove useful.

4-7 Bands — General

Most simple metals (Cu, Ag, Au, ..., Na, K, ..., and many more) and conventional superconductors contain conduction electrons that extend over many unit cells, and their wavefunctions can be approximated by Bloch waves. This leads to electronic bands in **k** space. However, the high-T_c materials are poorly conducting oxides, with low (for a metal) carrier concentrations, and some degree of two-dimensionality. For these various reasons, some have felt that extended-state Bloch functions may not be a good starting approximation to describe the behavior of the carriers in the normal, and superconducting, state of the high-T_c materials. The proposals to describe the normal state are complicated and we will not go into any detail but give a few references in the Notes. Rather, at an elementary level, we discuss a few of the areas that are discussed in the literature to at least understand the vocabulary. Then, in Sections 4-8 and 4-9, we return to conventional (one-electron) band theory calculations and comparison to experiment.

4-7a Fermi liquid — In a **Fermi gas**, or **free-electron gas**, electron-electron and electron-ion interactions are neglected. The positive ions are assumed to have a uniformly background density, which keeps the electrons in the crystal, and then the "free" electron behavior is only governed by the Pauli-exclusion principle, which leads to Fermi-Dirac statistics. The electrons have energies up to E_F corresponding to a maximum velocity v_F and wavevector k_F, so that $E_F = m \, v_F^2/2 = \hbar^2 \, k_F^2/2m$, and the Fermi surface is spherical. The Fermi gas model, with its assumption of independent electrons, has many remarkable successes in explaining the behavior of metals.

Landau (1957) investigated the electron-electron interactions and how they affect the independent electron approximating of the Fermi gas model. The results are usually called **Landau's Fermi-liquid theory**. Using subtle arguments (Notes), Landau arrived at some surprising conclusions. He showed that the large electron-electron does not charge the general E vs. **k** structure of the independent-electron model. Rather, the major effect of the interactions is that the effective mass of the electrons (now called quasi-particles because of their dressed character) increases by amounts up to ~50%. These values are of the order of those observed in many metals from electronic-specific heat data, for example. Further, the low-lying single-particle excitations of Fermi liquid are in one-to-one correspondence to those of the Fermi gas, only the mass is slightly changed from the bare mass (m) to an effective mass (m*).

Landau's Fermi-liquid theory gives a satisfying basis for the Cooper-pair problem and the BCS ground state. Within this theory, Cooper's problem can be understood as being a two-quasi-particle problem in a completely parallel way to being a two-particle problem in a Fermi gas. Further, the low-lying excitations of the Cooper problem and BCS ground state are similar in the Fermi gas or Fermi-liquid model. For high-T_c materials, understanding **normal-state properties** in terms of Fermi-liquid theory is an active area of research (Notes).

4-7b   Resonating-Valence-Band State — A full treatment of electron-electron repulsion effects in metals or insulators is difficult. A fruitful approach uses the **Hubbard model**. If one electron, of either spin, is on an atom, then its energy is E. However, if two electrons (of opposite spin) are in the same energy level on the same atom, the energy is 2E+U. It is assumed that the additional energy U comes from the Coulomb repulsion that occurs between two electrons on the same atom. Thus, this model replaces all of the complicated electron-electron repulsion terms with a single term, U, the **Hubbard energy**. To allow for some mobility of the electrons from one atom to another, a **site-hopping matrix element** t is defined.

If there is one valence electron from each atom, any sort of extended-state-band picture predicts a half-filled band and, therefore, metallic behavior. However, with the Hubbard model and a reasonably large U, an insulator is obtained, because it cost too much energy to have anything but a single electron on a single atom. In fact, the Hubbard model predicts an insulator that is antiferromagnetic with an exchange constant $J = t^2/U$.

Superconductivity in most of the cuprates occurs near a metal-insulator transition into an odd-electron insulator phase with peculiar magnetic properties. This has led Anderson to propose that this insulating phase might be the long-sought **resonating–valance–band (RVB)** state, first investigated in 1973. The RVB state is a new kind of state that is an alternative to an antiferromagnetic, $S=1/2$ state. When doped sufficiently to become a metal, the insulating state magnetic-singlet pairs become charge superconducting pairs. Thus, the pairing mechanism is predominantly electronic and magnetic. Another motivation for the RVB model is that for many of the high-T_c materials, metallic behavior is observed in the **ab** plane, but not along the **c** axis; rather, resistivity behavior similar to that in Fig. 4-3c is observed. On the other hand, at least for Y123, metallic behavior along the **c** axis is reproducibly obtained (see Fig. 4-3a and the Notes). The RVB formalism is complicated and beyond the scope of this book. Fortunately, the discussion between Anderson and Schrieffer (Physics Today, June 1991) helps in many aspects, and should not be missed. Also, the forthcoming book by Anderson (Notes) should be a good source.

4-7c Band Theory — Starting with the free electron idea and adding the atomic cores with their translational symmetry, E vs. **k** can be calculated, and this gives bands and what is usually called band theory. These calculations have now become quite sophisticated and the results described in Section 4-8 are calculated using the **local density approximation** or **LDA**, sometimes called **local density functional theory** or **LDFT**. This approach treats electron-electron correlation and exchange effects in terms of local values of the charge density and its derivatives.

These band calculations yield E vs. **k** over a wide energy range, not just near E_F. Thus, predictions of the excitations of the system can be compared to experiment in range of about 10 dV about E_F. This makes the band calculations more general on an energy scale than the Fermi-liquid theory, which presumably is only good for excitations near E_F. One-electron band results are discussed in Section 4-8.

4-7d Simple Two–Dimensional Bands — A structurally dominating aspect of the cuprates is the two-dimensional Cu-O planes. This and other physical results have led to consideration of these metals from a two-

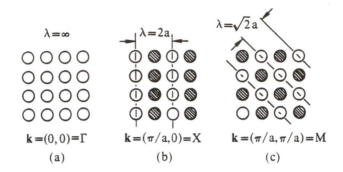

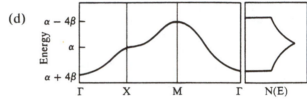

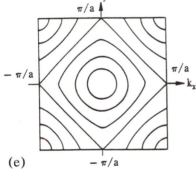

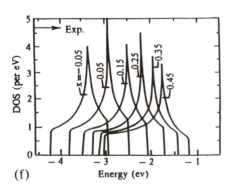

Fig. 4-7 (a) to (c) show s orbitals on a square lattice for **k** values as labeled. The white and shaded circles represent wavefunctions with positive and negative phase, respectively. (d) E vs. **k** around the first Brillouin zone. (e) The density of states vs. energy. (f) Constant energy contours in the first Brillouin zone showing the effects of increasing the number of carriers. A useful two-dimensional picture of E vs. **k** can be seen in A. Kitazawa, in Bednorz and Müller, Ed. (Bib.), where the saddle points are clear. (g) The mean field density of states (DOS) of a Cu-O plane model, optimized for 2-Bi(n=2), for a range of dopings, indicated by x. The horizontal bars denote the DOS at E_F; note E_F remains close to the van Hove singularity. The arrow on the left comes from the analysis of experimental data on 2-Bi(n=2). See D. M. Newns, P. C. Pattnaik, and C. C. Tsuei, Phys. Rev. B **43**, 3075 (1991).

dimensional point of view. In this section, some background basic physics of the two-dimensional problem is discussed.

Consider a two-dimensional square lattice of atoms, as in Fig. 4-7a. Take $\phi(\mathbf{r})$ as the eigenstate of an isolated atom, which we assume, for visualization purposes, is a normalized, nondegenerate s state with eigenvalue E_0. Using the **tight binding approximation**, assume that the overlap of $\phi(\mathbf{r})$ with its nearest neighbors is small, and that the atoms lie on a lattice of points given to be $\mathbf{t}_m$. When the electron is near $\mathbf{t}_m=0$, its eigenfunction is given approximately by $\phi(\mathbf{r})$, and when it is near the atom at lattice point $\mathbf{t}_m$, its wavefunction is approximately $\phi(\mathbf{r} - \mathbf{t}_m)$. Thus, the wavefunction for one electron in the crystal is

$$\psi_{\mathbf{k}}(\mathbf{r}) = \sum_m C_{\mathbf{k}m} \, \phi(\mathbf{r} - \mathbf{t}_m) \qquad (4-7a)$$

where the sum is over all of the lattice points. This wavefunction is a **linear combination of atomic orbitals (LCAO)**. As in standard tight binding, first-order perturbation theory is used and only self-energy terms from nearest neighbors are kept. Then, the constants,

$$\alpha = \, < \phi_m \, | \, H_{cry} \, | \, \phi_m >$$
$$\gamma = \, < \phi_k \, | \, H_{cry} \, | \, \phi_m > \qquad (4-7b)$$

are the self-energy and nearest-neighbor energies due to the crystal potential. the former case

For the two-dimensional problem under consideration, the tight-binding calculation yields an energy that depends on $\mathbf{k}$ as

$$E(\mathbf{k}) = \alpha + 2\gamma[\, \cos k_x a + \cos k_y a] \qquad (4-7c)$$

For s orbitals, the contributions for special $\mathbf{k}$ values are easy to draw, and Figs. 4-7a to 4-7c show some results. Energy (Eq. 4-7c) along the special lines and points in the Brillouin zone of the two-dimensional, square lattice are shown in Fig. 4-7d. The density of states is plotted in Fig. 4-7e showing a *logarithmic singularity* at half filling; this is also called a **van Hove singularity**. Figure 4-7f shows the constant-energy contours as the first-Brillouin zone is filled. Note the square Fermi surface at half filling, corresponding to the singularity in the density of states.

From Figs. 4-7d to 4-7f, several important observations can be made. First, for low filling of the band, or low E_F, the carriers are free-electron-like, with the conduction band centered at the Γ point (Fig. 4-7d). Second, if the band is more than half filled, the carriers are hole-like, with a a Fermi surface centered at the M point (Fig. 4-7d). Third, at half filling, the $E(\mathbf{k})$

are parallel to each other (Fig. 4-7f), which is called **nesting** or **nested bands**. There are four van Hove singularities at $k=(\pm \pi/a, 0)$ and $k= (0, \pm \pi/a)$; E vs. **k** has saddle points for these **k** values (see Fig. 4-7f caption), and this leads to the logarithmic singularity in the density of states (Fig. 4-7e).

Nesting could lead to a **charge density wave** (CDW) or a **spin density wave** (SDW). In the former case, the structure will also distort because of the Coulomb interaction between the electrons and the ions. Such a CDW usually causes an energy gap to open at the Fermi energy, lowering the energy of the entire electron system. This effect should lead to this half-filled electron system being a narrow gap insulator and not at all a metal. This effect, in general, can give rise to an incommensurate phase transition, since the periodicity is related to $2k_F$, the Fermi wave vector, which may have little to do with the repeat of the atomic structure. However, second neighbor interactions can alter the nesting of the Fermi surface.

For Figs. 4-7a to 4-7c, and the discussion, we have used s orbitals. These appear as bonding (Fig. 4-7a) and antibonding (Figs. 4-7b and 4-7c) states as in Section 4-2. However, p, or other orbitals, could just as well have been used, which also give rise to Van Hove singularities.

4-7e   More Advanced Two-Dimensional Bands — The two-dimensional bands discussed in the previous section have perfect nesting at half filling, which could lead to a CDW instability. However, this need not be taken too seriously, since next-nearest neighbor interactions, for example, will bow the square Fermi surface at half filling, decreasing the possibility of a phase transition. Thus, a logarithmic singularity may exist in the density of states, with the system remaining a metal.

In fact, even in conventional superconductors, the Fermi level near such density of states singularities have been considered (1967) as a possible enhancement mechanism for T_c in the conventional superconducting A15 compounds (Section 3-4). The high-T_c materials, with two-dimensional aspects to their structure and high values of T_c, have caused a considerable renewed interest in this field. We can not go into details, as many of the theoretical calculations are beyond the scope of this book; a few of the references are included in the Notes. However, we summarize some selected aspects of the results.

For the 2D problem, some recent theoretical approaches involve a **slave boson formalism** to calculate the bands, the density of states, and the position of the Fermi level as a function of doping. One very important result of these calculations is that the bands do not fill in the manner suggested by Fig. 4-7f, which is appropriate for rigid bands. Rather, as more holes are added to the system, they tend to go into the energetically low-lying oxygen states (Section 4-2b). This might be called charge rigidity of the Cu $3d^9$ system. In other words, *the Cu atom has a tendency to remain $3d^9$-independent of doping*. This can be seen in Fig. 4-7g, where each density of states (DOS) vs. energy corresponds to a different hole concentration. The DOS on the left is the least hole-doped and is the narrowest. The DOS on the right has the highest hole doping and is the widest, with the extra holes going into the planar oxygen orbitals, which appear as the foot on the low-energy side of the DOS.

Since the Fermi level lies near the $3d^9$ part of the band (Section 4-2b), roughly independent of doping, the paramagnetic susceptibility, as measured by the **Knight shift**, would be expected to remain mostly d-like. In fact, this is in agreement with NMR data. On the other hand, slave-boson calculations of angle-integrated photoemission (Section 4-9) show that the d-weight of the quasi-particle peaks should be ~40%, also in agreement with experiment.

A second important result of these calculations is that, since the added holes tend to be accommodated in the oxygen-like foot of the bands, the Fermi levels that lie in the Cu-like part of the band tend to remain near the singularities. This can be seen in Fig. 4-7g, where the doping is increased with increasing x values. The bands get wider, as discussed above, and the Fermi level remains very close to the van Hove singularity passing from just below it to just above it as the doping is increased over a wide range.

As might be expected, if the Fermi level lies close to a logarithmic singularity in the density of states, many unusual properties may be found in the material. A dramatic normal-state effect is in the quasi-particle lifetime τ. Modeling $1/\tau$ from electron-electron scattering yields an unusual result. For an ordinary metal, $1/\tau \propto (E - E_F)^2$, where $E - E_F$ is the energy separation of the excitation from the Fermi energy. For low-lying excitations, the scattering cross section from electron-electron scattering is small. However, near a van Hove singularity in the density of states, $1/\tau \propto (E - E_F)$ is found, leading to large cross sections for electron-electron scattering, which can become the dominant process controlling lifetime broadening.

One consequence of this stronger $1/\tau$ behavior is that the resistivity $\rho(T)$ becomes a linear function of temperature. An approximately linear ρ vs. T is found for optimal doping, which means the Fermi level is right at the van Hove singularity and the highest T_c is obtained. However, calculations also show that as one dopes away from optimal, T_c decreases and ρ vs. T shifts from being approximately linear to approximately quadratic. Such a shift in ρ vs. T has been recently experimentally observed in 2-Tl(n=1) (Notes).

Some of the unusual normal-state properties that should arise from the Fermi energy being near a van Hove singularity are briefly mentioned here. However, unusual superconducting properties should also arise; these are briefly discussed in Section 5-6h.

4-8 Band Theory

4-8a Introduction — Pre-1986, crystals as complicated as the high-T_c materials tended not to be considered by most solid-state physicists, a situation that has now changed. Even the electronic bands of the high-T_c materials have been calculated yielding results that may be called "spaghetti diagrams" because of the large number of fairly dispersionless electron energy levels vs. crystal momentum, $E(k)$, obtained. We focus our attention on the results for energies near the Fermi energy, E_F, since it is in these bands that are of immediate interest in understanding the conduction process and superconductivity.

Before discussing details and the some relevant experimental measurements, recall the basic assumptions that go into a band calculation. Band diagrams are essentially always for one-electron bands. In these calculations, the electron-electron repulsion is treated in an averaged manner. For electron orbitals whose extensions are large compared to the nearest neighbor distances, screening severely reduces the Coulomb-repulsive effect, making this a good approximation. Then, if considered at all, electron-electron correlation effects can be considered as perturbations. However, for the high-T_c materials, it has been argued by some that the electron-electron repulsion in these oxides is much larger than in ordinary metals; then the electrons are strongly correlated, so one-electron bands have little meaning. A Hubbard-type or resonating-valence-band model probably is a more sensible starting point to describe the electronic energies. Certainly, this point of view is supported by the fact that the one-election band calculations predict metallic behavior for all of the high-T_c type materials, even the ones that are *insulators*. On the other hand, for the cuprate *metals* (and superconductors), there is reasonable agreement between the one-election band-structure calculations and experimental results (discussed later). Thus, it appears that in the metals there are enough free carriers to provide significant screening so that E(**k**) may be reasonably well described by band-structure calculations using, for example, the **local–density approximation (LDA)**. Thus, we shall focus on the metals and their calculated one-electron bands of these high-T_c systems, and only mention the insulators in the Notes. However, as can be seen, there are some fundamentally different approaches to the behavior of the carriers in these materials. We are presenting just the simplest point of view, but one that is in reasonable agreement with experiments (Section 4-9) on the metals. In this section, one-electron band results are first considered. Then, the two-dimensional band results were discussed (Section 4-7d). This consideration leads to the van Hove singularity, which provides a sharp peak in the electronic density of states near E_F. Then some other electronic band considerations are discussed.

4-8b One–Electron Bands — The first Brillouin zone for a structure with a primitive-tetragonal (pt) lattice is shown in Fig. 4-8a with the special points and lines labeled in the conventional manner (Notes). This Brillouin zone is appropriate for Y123 and all of the 1-Tl(n) materials to the extent that they can be considered tetragonal (Chapter 3). The first-Brillouin zone for a structure with a primitive-orthorhombic lattice is shown in Fig. 4-8b, appropriate for orthorhombic Y123, for example. The corresponding Brillouin zone and labeling for structures with a body-centered tetragonal (bct) lattice, with $c/a > 1$, is shown in Fig. 4-8c. This Brillouin zone is appropriate for the 2-Tl(n) and 2-Bi(n) materials to the extent that they can be considered to be tetragonal (Chapter 3). The volume of the **irreducible wedge** outlined for the Brillouin zones of both tetragonal lattices (Figs. 4-8a and 4-8b) is $1/16$ of the volume of the first Brillouin zone. This factor corresponds to the 16-point-symmetry operations of the point group of the space group $(4/mmm - D_{4h})$ of both the primitive and the body-centered tetragonal lattices. In direct space, the corresponding volume of the unit cell is called the **asymmetric unit**. Using the point symmetry operations on either wedge yields the corresponding full unit cell. For orientational purposes, the point corresponding to the X point, but rotated 90° about the z axis, is often called the Y point; of course, the X and Y points are symmetry-related.

The first Brillouin zone in reciprocal space is obtained in the same way as the Wigner-Seitz cell in direct space and hence, in the extended zone scheme, both cells fill all space. This is shown in Fig. 4-8d, the first Brillouin zone of the bct lattice. Using Fig. 4-8d, to help visualize the extended zone scheme appropriate to Fig. 4-8c, we can understand an important point.

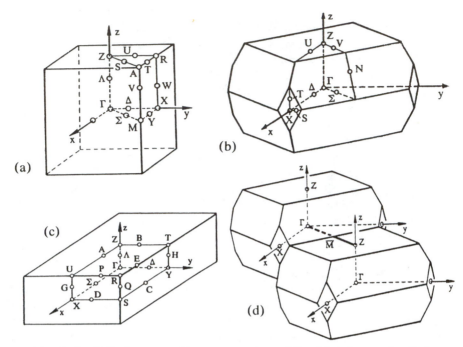

Fig. 4-8 The first Brillouin zones, with symmetry points and lines labeled conventionally for structures with: (a) a primitive-tetragonal (pt) lattice; (b) a body-centered tetragonal (bct) lattice, with c/a >1; (c) a primitive-orthorhombic lattice; (d) Stacking of the first Brillouin zones of the bct lattice.

Using Fig. 4-8b, start at Γ and move along the Σ line outside the first Brillouin zone; then one arrives at the Z point of the neighboring cell. Thus, in the *extended zone scheme*, and considering k in the (001) plane, the Z point will be encountered. For notational purposes, the point where the Γ-Σ line crosses the Brillouin zone, on the way to the Z point in the extended zone picture, is labeled $\overline{M}$ (Fig. 4-8d).

The calculated one-electron bands near E_F (taken as zero) are shown in Fig. 4-9a for 2-Bi(n=2). The short Γ-Z line on the right side of the figure shows $E(k)$ for bands along the [001] direction in reciprocal space. As seen, there is negligible band dispersion in this direction, as might be expected, since conduction is very small along c. Since our interest is primarily in conduction in the **ab** plane, we focus on the (001) plane in reciprocal space. The *extended-zone scheme* for the (001) plane is shown in Fig. 4-8d, where the Fermi surface, taken from $E(k)$ in Fig. 4-9a, can be seen in Fig. 4-9b. The **Fermi surface** is the intersection of E_F with the calculated $E(k)$.

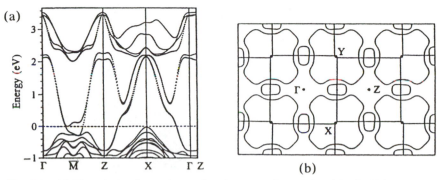

Fig. 4-9 (a) The calculated one-electron band structure for 2-Bi(n=2) in the high-symmetry directions. (b) Calculated Fermi surface in the (001) planes are shown using the extended zone scheme. In both parts of the figure, the labels Y and $\overline{M}$ are added for convenience of orientation. For details and references, see W. Pickett, Rev. Modern Phys. **61**, 433 (1989).

It is immediately obvious that the Fermi energy crosses some of the bands (Figs. 4-9a and 4-9b), hence the crystal should be a metal, at least in the **ab** plane. Crossings occur in the Γ-X and (equivalent Y) directions as well as in the Γ-$\overline{M}$ directions. Most of the bands below E_F arise for hybridized Cu-O orbitals, and the calculations indicate that the energy levels crossing E_F in the Γ-X direction arise primarily from these orbitals. The Bi-O bands lie primarily 2-4 eV above E_F. However, near $\overline{M}$ (between Γ and Z in the extended zone scheme), the Bi-O bands strongly disperse to lower energy (Fig. 4-9a) and form small, rounded rectangular electron pockets that arise primarily from these Bi-O hybridized states. This aspect of the calculation, namely, the existence of Bi-O like states near the Fermi energy, is an unexpected result.

4-9 Photoemission Spectroscopy

4-9a Introduction — Photons of sufficient energy, incident on a sample, can cause electrons to be emitted from the sample. A study of the energy of the emitted electrons is called **photoemission spectroscopy (PES)**. A schematic energy level diagram for a crystalline solid and a photoemission spectrum are shown in Fig. 4-10. The filled valence bands, just below the Fermi level, are shown crosshatched, and a much deeper, core level lies at an energy E_B below E_F. Radiation of energy hν impinges on the solid, and

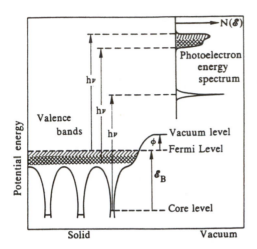

Fig. 4-10 (a) A schematic energy level diagram showing a photoemission spectrum from the valence bands as well as from a deeper (and hence sharper) core level. (b) The geometry, defining θ and ϕ, in a PES experiment.

electrons with kinetic energy E_{kin} are emitted as given by Einstein's (1905) photoelectron equation,

$$E_{kin} = h\nu - E_B - \phi \qquad (4 - 9a)$$

for the core level, where E_B is conventionally measured with respect to E_F ($E_B > 0$) and ϕ is the **work function** of the crystal (typically 2-5 eV). The number of electrons with kinetic energy greater than a fixed energy can be measured with a retarding potential on a grid between the sample and detector, allowing E_B to be obtained with the use of Eq. 4-9a. In the valence band energy region, the density of states $N(E)$ replaces E_B in Eq. 4-9a if **angle–integrated PES** is measured, that is, if electrons emitted at all angles are collected.

Depending on the energy of $h\nu$ and the use, PES has been called many things. When $h\nu$ corresponds to soft x-ray energies (~ 1 keV), the spectroscopy has been called **x-ray photoemission spectroscopy** (XPS) or **electron spectroscopy for chemical analysis** (ESCA). The early low-energy PES measurements were done with a He resonance lamp ($h\nu = 21.2$ eV) and the field was called **ultraviolet photoemission spectroscopy** (UPS). Now, many PES measurements are carried out using synchrotron radiation, which is a very intense source of radiation over an extremely wide energy range (Notes).

In PES experiments, the light can penetrate a considerable distance into the sample, but the 0-20 eV photo-emitted electrons are strongly scattered and absorbed in the crystal. The electrons collected in a PES experiment mostly come from a 0.5-2 nm surface region of the crystal. In fact,

PES has been effectively used to measure surface-specific electronic energy levels. With this surface sensitivity, care must be exercised that, indeed, properties related to the bulk are measured. Thus, PES experiments are carried out with the samples in an ultra-high vacuum (UHV) environment. UHV semiconductor or metal samples are usually "cleaned" by various complicated procedures such as cycles of low-energy ion bombardment and thermal annealing. Such procedures would not be appropriate for high-T_c cuprates oxides. Rather, the PES measurements are carried out on UHV, low-temperature cleaved crystals. Even with careful experimental procedures, care must be taken that the $E(\mathbf{k})$ results represent the bulk.

In **angle-resolved photoemission spectroscopy**, use is made of the fact that the component of momentum of the light and emitted electrons parallel to the surface is conserved in a PES process. The momentum perpendicular to the sample surface is not conserved, since the photoelectron transfers some perpendicular momentum to the crystal when escaping through the surface barrier. In general, the kinetic energy of an electron (T) is related to its momentum (p) by $T = p^2/2m$. Defining the angles as usual, the momentum parallel to the crystal surface is

$$p_\parallel = p \, \sin \theta = (2mT)^{\frac{1}{2}} \sin \theta \qquad (4-9b)$$

Therefore, the wave vector parallel to the crystal surface is

$$k_\parallel = p_\parallel / \hbar = (2mT/\hbar^2)^{\frac{1}{2}} \sin \theta \qquad (4-9c)$$

This k value is conserved in a PES experiment as long as the electron is not scattered when leaving the crystal. Further, by varying ϕ, which is measured with respect to the crystallographic axes, $\mathbf{k}$ in the plane can be determined.

4-9b PES 2-Bi(n=2) Results — 2-Bi(n=2) crystals are well suited for $E(\mathbf{k})$ measurements by angle-resolved PES. Relatively large crystals can be grown, which readily cleave in the ultra-high vacuum (UHV) environment required for PES experiments. It appears that the 2-Bi(n=2) crystals cleave between the two Bi-O planes (Fig. 3-1c), a result determined by x-ray PES measurements of the core levels. Hence, the body-centered tetragonal high-T_c structures, with an *even number of isolation planes* between the Cu-O planes, may be better candidates for PES experiments than the structures with the primitive-tetragonal type structures, such as Y123 (Chapter 3). For example, 2-Bi(n=2) crystals cleaved in UHV at 20 K have been cycled to just above T_c and back with no detectable changes in the PES spectra, while Y123 crystals treated similarly yield less reproducible results.

Thus, for a cleaved (001) plane of 2-Bi(n=2), by fixing the angle between the **c** axis and the **a** and **b** axes, the crystal momentum, **k**, in the plane is determined (Eq. 4-9c). Thus, with angle-resolved PES, E vs. **k**, or $E(\mathbf{k})$, in the **ab** plane can be measured. Also, the **k** value corresponding to the position where the electron bands cross the Fermi surface can be determined. For the present 2-Bi(n=2) PES results, the external angle resolution is $\pm 2°$, which for the 22 eV photons used corresponds to $\Delta k_{\parallel} = 0.075 \text{\AA}^{-1}$.

Experimental PES results are shown in Fig. 4-11a. The measurements (just above T_c, at 90 K) are along a line in **k** space that is parallel to the Γ-Y line at **k** values shown by the light dots in Fig. 4-11b. The data show that from at least 350 meV below E_F, the energy level band increases to higher energy. The data is fit by using a Lorentzian for the intrinsic spectrum plus a term linear in $E-E_F$ to account for the intensity at higher binding energy, multiplied by a Fermi-Dirac function, then broadened by the instrument function. The Fermi energy is determined independently by measuring a clear platinum foil in the same apparatus, which also reveals that the shape of the PES results near E_F for Pt and 2-Bi(n=2) are very similar. For **k** values corresponding to 12°, the Lorentzian peak of the PES response is cut off by the Fermi-Dirac function, which indicates that the band has crossed E_F.

From this type of measurement, $E(\mathbf{k})$ can be determined and some experimental results are shown in Fig. 4-11b along with calculated $E(\mathbf{k})$ results in the **k**-space direction. The crossing of E_F by the PES measured band is in excellent agreement with calculated one-electron band results. Along the Γ-Y line, the bands arise principally from Cu-O energy levels. Note (Fig. 4-11b) that $E(\mathbf{k})$ is nearly parabolic, so an experimental effective mass can be obtained; the result is $m^* \approx 2m$. On the other hand, the value calculated at this Fermi level crossing is $m^* \approx m$. This difference is apparent from the slopes of $E(\mathbf{k})$.

PES measurements along Γ-$\overline{M}$, and in the vicinity of $\overline{M}$, show that the measured band crossing of E_F is in good agreement with what is predicted by the calculations (Fig. 4-8). As mentioned, the calculated bands at E_F near $\overline{M}$ arise from Bi-O energy levels, and it was believed by many that such energy levels would not be near E_F because conduction in the **ab** plane would take place only in the Cu-O planes. There are some predominantly Cu-O bands just below E_F near $\overline{M}$; some have argued that these are actually the energy levels at E_F. In the $\overline{M}$ region, the experimental PES data of $E(\mathbf{k})$ is more complex than the nearly parabolic bands along the Γ-Y line

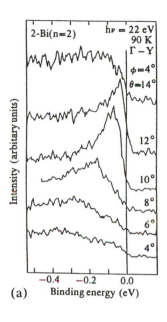

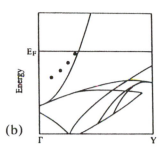

(b)

Fig. 4-11 (a) Angle-resolved PES energy distribution curves for several angles along the Γ-Y line in the Brillouin zone for 2-Bi(n=2). Different angles correspond to different **k** values (Eq. 4-9c). E_F is taken as 0 eV, 22 eV photons are used in the PES experiment, (b) The measured band dispersion and calculated $E(\mathbf{k})$ along the Γ-Y line. For details, see C. G. Olson, R. Liu, D. W. Lynch, R. S. List, A. J. Arko, B. W. Veal, Y. C. Chang, P. Z. Jiang, and A. P. Paulikas, Phys. Rev. B **42**, 381 (1990) and Solid State Commun. **76**, 411 (1990).

(Fig. 4-11a); thus, an effective mass cannot be estimated. This complex nature of $E(\mathbf{k})$ near $\overline{\mathrm{M}}$ is in qualitative agreement with the calculations.

The nature of the bands near $\overline{\mathrm{M}}$ were investigated in an interesting way. Less than ½ of a monolayer of gold was evaporated onto the 2-Bi(n=2) crystal in UHV. It should be expected that the Au would only interact with atoms in the outermost surface layer of the crystal. As mentioned, there is evidence that the Bi-O planes make up the surface layer of these crystals in UHV cleavage. After depositing the gold, the $E(\mathbf{k})$ results in the region of $\overline{\mathrm{M}}$ were drastically changed, while those along the Γ-Y line were not. From this, it was argued that the bands near $\overline{\mathrm{M}}$ indeed arise from the Bi-O bands, while those along the Γ-Y line arise from the Cu-O bands, which are below the surface of the crystal, hence are less effected by the gold overlayer.

4-9c PES Y123 Results — Reproducible PES results from Y123 have been more difficult to obtain because the crystal surface undergoes rapid changes (presumably due to oxygen loss) when subjected to UHV at room temperature. However, when the crystals are cleaved at low temperatures in UHV, they appear to be stable and show a large density of states at E_F. PES results on such crystals have been reported.

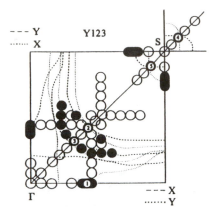

Fig. 4-12 Filled circles indicate experimentally determined points where the bands cross the Fermi surfaces in Y123-$O_{6.9}$. Dashed lines indicate the calculated Fermi surfaces. See the text for more details and see J. C. Campuzano, G. Jennings, M. Faiz, L. Beaulaigue, B. W. Veal, J. Z. Liu, A. P. Paulikas, K. Vandervoort, H. Claus, R. S. List, A. J. Arko, and R. J. Bartlett, Phys. Rev. Lett. **64**, 2308 (1990), and J. Yu, S. Massida, A. J. Freeman, and D. D. Koelling, Phys. Lett. A **122**, 203 (1987).

PES results on stable Y123 crystals are shown in Fig. 4-12. Since the samples are twinned, it is not possible to distinguish between the Γ-X and Γ-Y directions, so the theoretical curves and experimental data shown in Fig. 4-12 are reflected about the Γ-S direction (Fig. 4-7b). The circles in Fig. 4-12 are points in the Brillouin zone where the measured energy distribution curves have been taken. The filled circles indicate that E(**k**) crosses the Fermi level. (Open circles indicate no Fermi-level crossing.) The dashed lines are calculated Fermi surfaces. The crossing labeled 1 in Fig. 4-12, which occurs at $(0\ 0,\ 0.35)\text{Å}^{-1}$, is part of the calculated Y-centered hole Fermi surface. The band crossings labeled 2 and 3 are part of an S-centered hole Fermi surface in the second-hole Fermi surface in the second Brillouin zone. The crossings labeled 4 and 5 are part of a hole pocket around S. In general, good agreement is found between the one-electron band calculation and the PES experiments with respect to the Fermi surface.

4-9d PES Summary — At present, the most extensive and consistent PES measurements have been on 2-Bi(n=2). This material can be grown in large crystals, cleaves easily in the **ab** plane, and appears to give reproducible results. Probably the most general conclusion that can be drawn from the measurements is that in the normal state, near E_F, the material looks like an ordinary metal. Further, the measured E(**k**) are consistent with the one-electron calculations. The more recent PES results on Y123 further substantiate these conclusions.

On the other hand, there are certain discrepancies that can be focused upon. First, the measured effective mass along the Γ-Y direction in 2-Bi(n=2) is about two times the calculated value. The lifetime of the electrons at energy E, in the range below E_F that has been measured, is linear in E-E_F rather than varying as (E-E_F)^2 as expected from standard band theory. We have not discussed the lifetime results, but they come from fitting of the experimental spectra. At present, it is not clear how serious the disagreements should be taken.

PES measurements on other high-T_c superconductors have also been performed. The results are at present less extensive and, because of experimental crystal problems, less reliable. However, PES appears to be an excellent way to probe the normal state bands. PES will be discussed again in Section 5-4a in connection with their use in measuring the superconducting energy gap in different k directions.

Problems

1. Sommerfeld formula — Derive Eq. 4-4a using the free electron approximation.

2. Two–carrier Hall effect — Derive Eq. 4-4b assuming both electrons and holes are present in parabolic bands. (Hint: both this and the previous problem are discussed in most introductory solid-state physics books.)

3. PES — (a) For a 22 eV photon impinging on a crystal with an angle 12° to the normal, calculate the parallel k value. Compare this value to: (b) the Brillouin zone edge k value in the X direction of 2-Bi(n=2), assuming that this material is tetragonal, and (c) the k resolution quoted in the text.

4. Molecular orbitals — Compare the calculations in Section 4-2 with the results of L. F. Mattheiss, Phys. Rev. Lett. **58**, 1028 (1987). Note where E_F crosses the calculated bands that arise from the antibonding orbitals that we discussed. Are all of the band-E_F crossings discussed in this paper?

Fig. 5-1 T_c vs. the year of discovery of superconductivity in a few of the oxide superconductors. The notation for the high-T_c materials is in Table 3-1. Nb_3Ge is included as a fiducial marker.

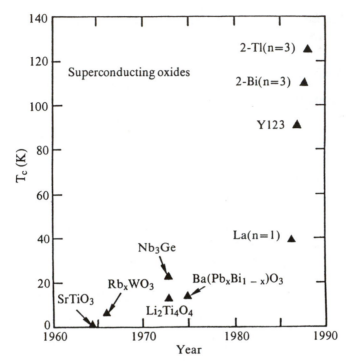

Chapter 5

Superconducting Properties

What is a prophet? Someone who utters one truth in a flock of lies,
if he is lucky, and if he is not lucky everyone forgets.

Euripides, "Iphigeneia at Aulis"

Here, we discuss high-T_c superconductors in a parallel manner to Chapter 2 on **conventional superconductors** (i.e., non Cu-O planar ones). The basic scales (energy, length, and time) are established. The energy scale is set by the transition temperature, T_c, about 10 times larger than found in conventional superconductors, and is intimately associated with the superconducting energy gap.

From the highly anisotropic crystal structures (Chapters 1 and 3), we should expect that superconducting length scales may reflect this anisotropy. The two important length scales are the **coherence length**, a measure of the spatial extent of superconductivity, and the **magnetic penetration depth**, a measure of the extent of penetration of an external magnetic field. These lengths are closely related to the critical magnetic field, H_{c1} and H_{c2}, values and anisotropy is found in these lengths and fields. Finally, the time scale is discussed in Chapter 6, where steady-state phenomena as well as nonequilibrium phenomena are considered.

5-1 T_c Values

The most impressive property of "high-T_c superconductors" is their high values of T_c. Before 1986, the highest T_c was ≈ 23.2 K for Nb_3Ge, and it was felt that if this value were surpassed, it would only be by a degree or two. Now many high-T_c materials have $T_c > 77$ K (the boiling point of liquid nitrogen), as shown in Fig. 5-1. Thus, the energy scale for these materials

is ~10 times that of conventional superconductors. Thus, in the superconductor phase, the energy to overcome vortex pinning (Section 6-2b), for
example, is much larger near T_c, and $(T_c/\Theta_D)^2$ is much larger than in the
conventional superconductors. T_c is even close to the Debye temperature,
which for the cuprates is ~300-450 K. (Compare to Table 2-2.)

The high-T_c materials have a common structural link in that they all
contain one or more immediately adjacent Cu-O planes in the unit cell.
These planes are perpendicular to the **c** axis and hence parallel to the **ab**
planes as discussed in Chapters 1 and 3. The larger the number of immediately adjacent Cu-O planes (larger n, Table 3-1), the higher the T_c, at
least up to n=3 or 4. For several of these families, materials with larger n
values have been synthesized, but T_c has not continued to increase. However, the n value is only part of the story. The material also must be a metal
and the number of carriers per Cu atom may be important. It is possible that
for larger-n materials there is less control of the stoichiometry and doping.
It has also been suggested that for n≥3, the charge carriers may reside primarily on the outer two planes of the n immediately adjacent planes. Thus,
adding more Cu-O planes is not helping but, to the contrary, only separating
these outer two planes from each other. However, it is also possible that a
better handle on doping the higher-n materials is all that is required. Considerable effort is being expended to "break the 125 K barrier."

For all of the high-T_c materials, T_c varies with doping, and Fig. 4-3c
shows the results for La(n=1). For x≳0.06, it is a metal and a superconductor, with T_c peaking at about 38 K for x=0.15. However, for x≳0.26,
although La(n=1) remains a metal, it is no longer a superconductor. In fact,
for x ≳0.26, it is a "better" metal in that the room temperature electrical
conductivity is larger than for smaller x values; this result is not understood.
The data in Fig. 4-3c appear to settle another controversial point. It had
been thought that La(n=1) had to be orthorhombic to be superconducting;
this appears not to be the case.

For $YBa_2Cu_3O_{7-\delta}$, the variation of T_c with doping (δ) is shown in
Fig. 4-4b. For $\delta \approx 0$, T_c is 93 K and it essentially maintains this value for a
composition range of approximately $O_{7.0}$ to $O_{6.8}$. For a smaller oxygen
content, T_c decreases, but note that for a range of δ, there is a "60 K plateau." The value of δ for which this material is no longer a metal varies with
sample preparation, as do both the 93 and 60 K plateaus. However, at
about $O_{6.35}$, there is a metal-insulator phase transition; Y123 is tetragonal
and becomes antiferromagnetic with a Néel temperature that is also shown
in Fig. 4-4b.

For most of the high-T_c superconductors (Table 3-1), T_c can be varied by changing the doping. This can be done by subjugating the crystals to excess oxygen pressure or a reducing atmosphere (at elevated temperatures). Or the doping can be changed by replacing some of the Y^{3+} by Ca^{2+}; both of these ions usually occupy positions between the immediately adjacent Cu-O planes. The T_c vs. doping (carrier concentration) curve for most of the high-T_c materials is a bell-shaped curve similar to those shown in Figs. 4-4c and 4-5c. Thus, for each high-T_c material, there appears to be an optimal doping for the highest T_c. Also, most of these materials can be doped so that they are insulators, and many can also be doped so that they become non-superconducting metals.

5-2 Cooper Pairs and BCS

5-2a Introduction — Fröhlich (1950) and others showed that the electron-phonon interaction could, under certain conditions, lead to an attractive interaction between electrons. This phonon-mediated electron-pairing interaction is attractive for electrons with energies within a characteristic phonon energy $\sim\hbar\omega_D$ of the Fermi surface. Using this idea, Cooper (1956) considered two electrons restricted to being outside of a filled Fermi sphere (filled to energy E_F) and not interacting with any other electrons. If these two electrons have an arbitrarily small attractive interaction for one another, then they will form a bound state with an energy lower than $2E_F$. This bound state is an orbital s-state in the relative position vector $\mathbf{r}$ implying that the wavefunction is even, under spatial interchange of the electrons, so the electrons must be in a singlet-spin state. The lowest energy of this pair of electrons occurs when its total crystal momentum $\mathbf{K}$ is zero. This discussion assumes a non-interacting Fermi gas, so it was unclear that such bound states would exist when interactions with the other electrons are considered. What was clear is that the free-electron Fermi gas may be unstable against the formation of such pairs. Thus, this interesting, insightful result, while not a model of the superconducting ground state, was an important clue.

With this insight, Bardeen, Cooper, and Schrieffer in the next year (1957) constructed a many-body generalization of the Cooper pair; this is the famous BCS model for superconductivity. The BCS many-body ground-state wavefunction for the many electrons is an antisymmetrized product of identical, pair wavefunctions, where each pair wavefunction has a total crystal momentum of zero and a total spin of zero. Above the ground-state energy, there is an energy gap, the superconducting gap $2\Delta(T)$, before one reaches the excitation spectrum. Thus, the BCS ground-state wavefunction is stable against single-particle excitations. With their ground- and excited-state wavefunctions, BCS were able to calculate essentially all of the equilibrium properties of a superconductor (Section 2-5).

5-2b   Paired Electrons? — Of course, in normal metals, since the charge carriers are fermions, they obey the Pauli-exclusion principle and, hence, Fermi-Dirac statistics. Thus, there can only be two electrons (with opposite spin) in any one orbital-quantum state at a time. Generally, it is felt that for superconductivity to occur, whether it be BCS or any other sort of superconductivity, the ground-state wavefunction must somehow be a paired state of the type discussed above. A test for a paired state is described below.

In conventional superconductors, by measuring the magnetic flux, Φ, trapped in hollow superconducting cylinders (1962), it was found that this flux is an integral multiple of a fundamental unit, the fluxoid quantum Φ_0, (Section 2-4d) such that

$$\Phi = n\ (hc/2e) = n\ \Phi_0 \qquad\qquad (5-2a)$$

where n is any integer. The factor 2 in the denominator shows that the superconducting ground state is composed of paired electrons. This was especially interesting because London considered the flux quantization problem in 1935 and arrived at the quantization, but had it in multiplies of hc/e, not appreciating the pairing in the superconducting ground state. Since the first experiments in conventional superconductors, other results always show paired electrons as the superconducting charge carriers. This suggests a critical question for the high-T_c materials: do we have paired electrons, or something completely different?

Early in 1987, experiments were performed on high-T_c materials to determine if the superconducting state consisted of paired electrons. A result from one of these experiments is shown in Fig. 5-2a. A ring (~10 mm and 5 mm outer and inner diameter, respectively) was formed from a Y123 sintered pellet and exposed to electromagnetic noise below T_c. The magnetic flux through this ring was measured (by a weakly coupled SQUID magnetometer) as a function of time. The quantized nature of the flux passing in and out of the ring can be seen in Fig. 5-2a. In a separate experiment, the magnitude of the flux was calibrated with a long solenoid passing through the ring. The experimental value for the fluxoid quantum

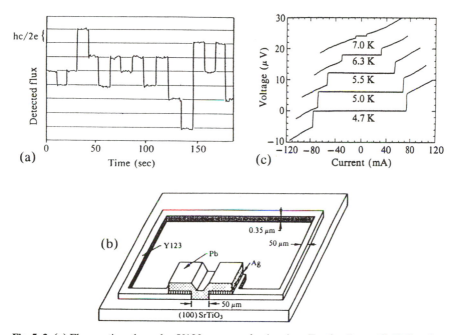

Fig. 5-2 (a) Flux vs. time through a Y123 superconducting ring. For details, see C. E. Gough, M. S. Colciough, E. M. Forgan, R. G. Jordan, M. Keene, C. M. Muirhead, A. I. M. Rae, N. Thomas, J. S. Abell, and S. Sutton, Nature **326**, 855 (1987). (b) Schematic diagram of the composite high-T_c and conventional superconductor loop. The gap in the ring patterned in the epitaxial c-axis oriented film is closed by an evaporated lead film. The loop area is typically a few mm². (c) Current-voltage characteristics of the Y123/Pb/Y123 structure at different temperatures. The temperature dependence of the critical current, approximately $1-(T/T_c)^2$, is typical for a type-I superconductor, indicating that the critical current of the structure is determined by lead. For details, see R. Gross, P. Chaudhari, A. Gupta, and G. Koren, Physica C **166**, 277 (1990); M. N. Keene, T. J. Jackson, and C. E. Gough, Nature **340**, 210 (1989); G. T. Lee, J. P. Collman, and W. A. Little, Physica C **161**, 195 (1989).

was $\Phi_0=(0.97\pm0.04)(hc/2e)$ as indicated in the figure. Using this technique (fluxoid quantization in a ring), as well as results from Little-Parks oscillations, and Shapiro steps (Section 6-7), it has been demonstrated that high-T_c superconducting carriers consist of paired electrons, and not something more complex. (See the Notes.)

5-2c Spin Singlet or Triplet Pairing? — In free space, a wavefunction of two electrons can be written as a product of a space and spin part,

$$\Psi(\mathbf{r}_1, s_1, \mathbf{r}_2, s_2) = \psi(\mathbf{r}_1, \mathbf{r}_2)\, \chi(s_1, s_2) \qquad (5-2b)$$

where $\mathbf{r}$ and s are the spatial and spin coordinates.

The total wavefunction for a pair of fermions (each $s=\frac{1}{2}$) *is antisymmetric with respect to an exchange of the two particles.* Thus, from elementary quantum mechanics, the total wavefunction (with total spin S) is either a spin triplet (S=1) or a spin singlet (S=0) and can be written either as

$$\begin{aligned}
\Psi_A^{S=1} &= \psi_{odd}(\mathbf{r}_1, \mathbf{r}_2)\, \chi_{even}^{S=1} \\
\Psi_A^{S=0} &= \psi_{even}(\mathbf{r}_1, \mathbf{r}_2)\, \chi_{odd}^{S=0}
\end{aligned} \qquad (5-2c)$$

where $\psi_{odd}(\mathbf{r}_2, \mathbf{r}_1) = -\psi_{odd}(\mathbf{r}_1, \mathbf{r}_2)$ and $\psi_{even}(\mathbf{r}_2, \mathbf{r}_1) = \psi_{even}(\mathbf{r}_1, \mathbf{r}_2)$. Letting α and β be spin up and spin down of an electron with respect to a given axis of quantization, and the numbers 1 and 2 refer to the two electrons as before, then the spin part of the wavefunction has the familiar form,

$$\chi_{even}^{S=1} = \begin{cases} \alpha(1)\alpha(2) \\ [\alpha(1)\beta(2) + \beta(1)\alpha(2)]/\sqrt{2} \\ \beta(1)\beta(2) \end{cases} \qquad (5-2d)$$

$$\chi_{odd}^{S=0} = \quad [\alpha(1)\beta(2) - \beta(1)\alpha(2)]/\sqrt{2}$$

Thus, it can be seen that when electrons 1 and 2 are interchanged, the S=0 wavefunction has an antisymmetric spin part so must have a symmetric space part. The S=1 wavefunction has the opposite (i.e., symmetric spin part and antisymmetric space part). Thus, for the spin-singlet wavefunction (S=0), the orbital part of the total wavefunction must have s-state, d-state, ... orbital angular momentum. Similarly, the spin-triplet (S=1) must have p-state, f-state, ... orbital angular momentum.

The BCS pairs are spin singlets with s-state orbital angular momentum. This is often called simply s-state or s-wave pairing, the rest being implied. Most conventional superconductors appear to have s-state pairing. However, the heavy-electron metals (Section 2-8c), at least UPt_3, appear to have d-state pairing. While in the conventional BCS model, the attractive interaction arises from the electron-phonon interaction, many non-phonon pairing mechanisms have been suggested for the high-T_c materials. Spin-fluctuation exchange mechanisms or mechanisms based on large on-site Coulomb repulsions tend to give d- or p-state pairing, but much more work remains to be done. Theories of this type are thought to explain the unconventional pairing in UPt_3 and ^3He (Section 2-9e).

5-2d Symmetry of Electron Pairs — Results of Josephson tunneling experiments are an argument for the paired electrons in high-T_c superconductors being in a spin-singlet s state. It was pointed out (1966) that Josephson-type tunneling should not be possible between paired electrons in two different superconductors unless the states have the same symmetry. If they were in different angular momentum states, they would have to absorb or emit angular momentum at the junction (by flipping an electron spin or generating a spin wave), which would cause dissipation. Since the Josephson tunneling experiment was between Y123 and a Pb/Sn point contact (the latter being an ordinary s-wave superconductor),sa then Y123 should also have paired electrons in an s state.

Other experiments that attempt to address the pairing-symmetry problem are measurements of the temperature dependence of the penetration depth λ (Section 5-7b), PES (Section 5-4a), and experiments that probe the single-particle tunneling from a conventional s-wave BCS superconductor into a high-T_c superconductor. Figure 5-2b shows the geometry of one of the latter type of tunneling experiments, where a c-axis-oriented Y123 film (Section 6-3a) is patterned by laser ablation. The silver provides a very low electrical resistance contact over the large area between Pb and Y123 so that essentially all of the current is along the c axis of Y123 (which is perpendicular to the $SrTiO_3$ substrate). The current path in the **ab** plane direction is "dead" due to the patterning step. Relatively large supercurrents have been observed in these Pb-Y123 rings. From these results, it is concluded that the superconducting wavefunction has phase coherence around the entire ring, including the interfaces.

In experiments such as this Pb-Y123 experiment, exactly what can happen at the interface is always important, in fact critical. It has been argued (see Notes) that to couple an s-wave to a p-wave-like order parameter, magnetic scattering at the interface is required. However, these materials and junctions supposedly are free from such impurities. It is more difficult to completely rule out a d-wave order parameter. However, since the current is flowing in the c direction in Y123, the positive and negative lobes of a d-wave order parameter should be zero when averaged over the s-wave order parameter of Pb. Thus, the large magnitude of the supercurrent appears to favor mostly s-state pairing in Y123. In agreement, the temperature dependence of the penetration depth (Section 5-7b) and PES experiments favor s-wave pairing and, within experimental error, allow only a small amount of d-wave pairing at most.

Although Josephson tunneling and conventional tunneling experiments appear to indicate s-wave pairing in Y123, many specialists in these areas are less convinced. They feel that "anything may happen at the junctions." There may be several angstroms of insulator at the junction interface and, in the insulator state, there could be magnetic Cu atoms available to cause electron-spin flips. Perhaps a conservative statement is that the tunneling results discussed here are consistent with a BCS-like singlet-spin s state, but, as yet, they are not totally convincing until the actual junctions are better understood. There are, however, other experiments that indicate Y123 has s-state pairing (Sections 5-7b and 5-7d).

One of the basic predictions of the (s-wave) BCS theory is the presence of a superconducting gap in the allowed energy states. This means that on both sides of the Fermi energy, there are no allowed states for a certain range of $\mathbf{k}$ values. This gap in the density of states leads to exponential activation energies for many physical processes. On the other hand, for non-s-wave paired superconductors, the superconducting band gap goes to zero along lines, or at points, in $\mathbf{k}$ space. This results in the density of states varying linearly or quadratically with energy. This has a consequence that many measured quantities vary in a power-law manner as $T \to 0$, while for the s-wave case, measured quantities vary exponentially as $T \to 0$.

Experiments on heavy-electron materials (Section 2-8c) show temperature-dependent, power-law behavior for many processes, indicating that the energy gap vanishes at points or lines on the Fermi surface. Thus, these superconductors are not s-wave, but d-wave or p-wave superconductors. Non-s-wave superconductors are sometimes called **unconventional superconductors**. The presence of two distinct superconducting transitions in UPt_3 and $(U,Th)Be_{13}$ is also compelling evidence for unconventional superconductivity in these materials.

Summary — For the high-T_c superconductors, essentially all experimental evidence indicates that there are no zeros in the superconducting gap function at any $\mathbf{k}$. Thus, these materials are s-state superconductors. Further, the gap appears to be 5-7k_BT_c in the **ab** plane and 3-4k_BT_c along the **c** axis. Thus, the superconductivity gap has considerable anisotropy compared to conventional materials (Section 2-6e). These results are obtained from experiments described in this section as well as other experiments discussed later in this chapter.

5-3 BCS Superconductors?

Are the high-T_c materials BCS superconductors? The answer depends on what is meant by the phrase BCS superconductor. The term "BCS superconductor" might be applied only to superconductors that fit the assumptions made in the BCS paper (1957). There it was assumed that the attractive electron-pairing interaction is due to electron-phonon coupling, that the pairs are weakly coupled compared to average phonon energies (called **weak-coupled BCS**), and that the pairs are in a spin-singlet s-state, which implies that the superconducting energy gap is isotropic. The extension to strong (isotropic) electron-phonon coupling is fairly straightforward, and is often called **strong-coupled BCS** (Section 2-6).

However, for these high-T_c materials, the electron pairing mechanism is not yet clear. There is a possibility that they are strong-coupled BCS. The high-T_c materials typically have Debye temperatures ~300 K to 450 K (Section 5-4c), although due to the many optical-phonon branches, the phonon density of states is not Debye-like. Nevertheless, these Θ_D values are similar to those of conventional superconductors (Tables 2-2 and 2-4). Thus, $(T_c/\Theta_D)^2$ is very much larger in high-T_c superconductors than in conventional ones; these large ratios point to strong coupling, if phonons are responsible for the electron pairing. However, some experiments are consistent with weak-coupled BCS. A case has been made that the strong unscreened electron-electron Coulomb interaction expected in these oxides is weakened when two-band models are considered. This weakening, along with a plausible coupling to high-energy optical phonons, might result in $T_c > 100$ K values.

Although these materials show electron-pairing superconductivity, it is possible that the pairing interaction may not be phonon-mediated. Antiferromagnetic magnons, excitons, and other mixed-coupling mechanisms have also been mentioned. Phonon-mediated superconductivity with some sort of a booster to increase T_c is also a possibility.

The high-T_c superconductors show many unusual effects. There is evidence for a large difference between the superconducting gaps in the **c** and **ab** directions. This anisotropy could be associated with the highly anisotropic structures found in these planar Cu-O materials. Within the **ab** plane, the gap appears to be closer to isotropic (Sections 5-4a and 5-7d). States within the superconducting gap have been discussed, but some of the experiments that indicate such states have been shown to be inappropriate (Section 5-4e).

It is possibly appropriate to use the term "BCS superconductor" for any pairing superconductor, because essentially all aspects of the basic BCS equations for the superconducting condensed state follow with a coupling of non-phonon origin. One needs only some other boson to cause "boson-mediated" superconductivity for pairs of electrons within a certain energy range about E_F. We shall not argue the merits of the use of the terms, but merely alert the reader to the semantic problem.

5-4 Superconducting Energy Gap and Other Properties

An important BCS prediction is the existence of a temperature-dependent energy gap $2\Delta(T)$ for the elementary excitations in the super-conducting state (Section 2-5c). There are no allowed states within this gap (Figs. 2-6a and 2-6b) and, for conventional superconductors, there is ample experimental evidence for this result. For weakly coupled BCS superconductors, the energy gap at 0 K is related to T_c by

$$\frac{2\Delta(0)}{k_B T_c} = 3.52 \qquad\qquad (5 - 4a)$$

as discussed in Section 2-5c. For the strong-coupled superconductors Pb and Hg, this ratio increases to 4.3 and 4.6. It is natural to ask if the high-T_c crystal has an energy gap, and, if so; what is its magnitude, does it vary in the different spatial directions (i.e., is it anisotropic), and are there any allowed states in the gap? For example, the resonance-valence-bond (RVB) theory tends to predict that states are allowed in the gap (i.e., there really is no gap in the superconducting state). In conventional superconductors, there is evidence for small anisotropies in the gap (Section 2-5e).

5-4a PES Results — Perhaps the least controversial measurements of the superconducting energy gap in high-T_c crystals come from **photoelectron spectroscopy**(PES) measurements of 2-Bi(n=2). Normal state PES measurements of this material were discussed in Section 4-7. Not only can the superconducting energy gap be measured, but because of the **k**-resolving capability of angle-resolved PES, gaps in different **k** directions can be determined. For 2-Bi(n=2), only the anisotropy in the **ab** plane has been investigated by PES because of the easy **ab** plane cleavage in vacuum.

Conceptually, the experimental PES procedure is straightforward. For a given direction in **k** space, for example along the Γ-X line (Fig. 4-8b), PES data is used to determine where the band crosses E_F. In the BCS the-ory, a superconducting energy gap, centered at E_F, should open (Fig. 2-6a) below T_c. Then, where the bands cross E_F, PES data is taken above and below T_c. For 2-Bi(n=2) crystals (T_c=82 K), PES data are shown in Fig. 5-3a at 90 K and 20 K. Due to the opening of a gap, the density of states is pushed away from E_F and piled up at higher and lower binding energies

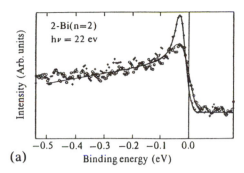

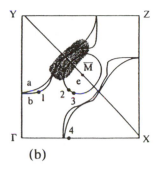

(a)   (b)

Fig. 5-3 (a) PES data (circles for T= 90 K and crosses for T= 20 K) for the point in k-space labeled 1 in (b) of this figure. The lines are the fits. (b) A calculated Fermi surface of 2-Bi(n=2) with four points labeled (1, 2, 3, 4) at which angle-resolved PES measurements were taken to determine $2\Delta(T)$. For details, see C. G. Olson, R. Liu, D. W. Lynch, R. S. List, A. J. Arko, B. W. Veal, Y. C. Chang, P. Z. Jiang, and A. P. Paulikas, Solid State Commun. **76**, 411 (1990).

(Figs. 2-6a and 2-6b). Of course, quantitative values of $2\Delta(T)$ require a detailed fit of the data. As discussed in Section 4-7, the data were modeled with a Lorentzian curve representing the intrinsic spectrum, plus a term linear in $(E - E_F)$ (also see Section 2-8b), multiplied by a Fermi-Dirac function, and then broadened by the instrument function. Below T_c, the line shape was also multiplied by a BCS density of states with the gap value taken as a parameter. The data and fit shown in Fig. 5-3a are for the point in reciprocal space labeled 1 in Fig. 5-3b. Of course, considerable modeling of the observed PES data is required to obtain 2Δ, and even a gap (with no allowed states) is assumed. Nevertheless, the fit, over a broad binding energy interval, is good.

These PES measurements were repeated at different points (labeled 1, 2, 3, and 4 in Fig. 5-3b) in the (001) plane of the Brillouin zone. These four points cover regions where bands due to Cu-O are dominant (near the Γ-X and Γ-Y lines) and where bands due to Bi-O are dominant (near $\overline{M}$), as discussed in Section 4-7. At these points in **k** space, the same BCS-type superconducting gap was obtained. The value found at 20 K was $2\Delta=48\pm10$ meV. Thus, taking $T_c=82$ K, $2\Delta(0)/k_BT_c = 6.8$.

These PES experiments imply several important results. First, fairly independent of modeling (except assuming a gap), the superconducting energy gap is isotropic in the **ab** plane within the accuracy of the measurements. This result argues for s-wave spin pairing within this plane, with at most a small amount of d-wave spin pairing, since d-waves would yield nodes within the plane. Second, the gap is larger than predicted by weak-coupled BCS (Eq. 5-4a). Third, by carrying out these PES measurements as a function of temperature, $2\Delta(T)$ can be determined. Some very preliminary results indicate a temperature dependence of the gap that is similar to BCS.

However, true to the high-T_c tradition, as soon as careful, fairly clear experimental results are published and generally accepted, conflicting data are also published. Very recent PES measurements have reexamined the gap anisotropy and now suggest that the gap in the ΓX direction may be smaller than that in the $\Gamma \overline{M}$ direction. The PES line shape in the $\Gamma \overline{M}$ direction is complicated, since the bands are not free-electron-like. We shall have to await the resolution of these measurements.

5-4b Tunneling Spectroscopy — For conventional superconductors, single-particle tunneling techniques have been very successful in determining the temperature dependence of the superconducting gap as well as confirming the electron-phonon mechanism (Section 2-7). Thus, it is natural to try similar methods in the high-T_c materials. However, the results are controversial and only recently is a consensus beginning to develop as to the meaning of tunneling measurements.

Experimentally, tunneling measurements in high-T_c superconductors have been performed using a range of approaches, as for conventional materials. These include:

(a) vacuum tunneling from a metal tip, through a vacuum ($\sim$10-100 Å), into the superconductor;

(b) point-contact tunneling, where a metal point is pushed into the superconductor;

(c) break-junction tunneling, where the sample is broken, pushed together, and tunneling between the pieces is measured;

(d) planar-junction tunneling, where a metal is evaporated onto the superconductor, with a thin (perhaps natural) oxide layer in between.

Figures 5-4a and 5-4b show good tunneling data in Y123 crystals and films at low temperatures by the point-contact technique. Unfortunately, not all of the tunneling data are so clear, and as T approaches T_c, it becomes more broadened with increased background conductance.

Point-contact measurements seem to provide some of the better data, but then the role played by gap anisotropy is not clear. If the metal tip is pushed into the crystal in the **c** direction, is the measured superconducting gap along the **c** axis or in the **ab** plane?

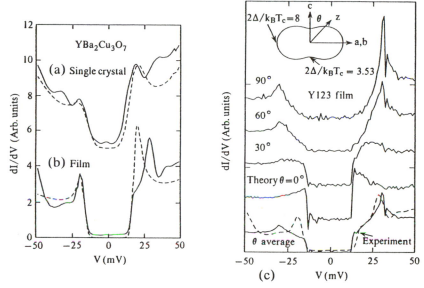

Fig. 5-4 Y123 point-contact tunneling data for tunneling into: (a) the **c** axis of a single crystal, (b) the **c** axis of an oriented film. The data are the solid lines, and the dashed lines are theoretical fits to a Gaussian distribution of gaps allowing for energy-dependent tunneling barrier probabilities. (c) Theoretical modeling of the effects of the superconducting gap anisotropy (shown at the top) on the tunneling conductance vs. applied voltage. For details, see the article by Kirtley referenced in the Notes.

It has been argued that the tunneling current should be the sum of contributions from carrier tunneling in all directions. In the simplest case, an anisotropic superconducting gap $\Delta(E, \mathbf{k})$ might be modified so that BCS density of states may be given by

$$N(E, \mathbf{k}) = \frac{|E - E_F|}{[(E - E_F)^2 - \Delta(E, \mathbf{k})^2]^{\frac{1}{2}}} \qquad (5 - 4b)$$

where $\qquad \Delta(E, \mathbf{k}) = \Delta_{ab} \sin^2\theta + \Delta_c \cos^2\theta$

and θ is the angle between **c** and **k**. Therefore, the experimental conduction vs. voltage should have an onset for carriers tunneling across the **c**-axis gap (taken as the lower energy gap) and then peak for carriers tunneling across the **ab**-plane gap (the higher energy gap). The latter has more tunneling carriers due to the larger solid angle subtended. Another problem in interpretation of experimental data is that the tunneling barrier heights may be low compared to the superconducting gaps. The upper part of Fig. 5-4c shows the angular dependence of the superconducting gap of Eq. 5-4b, assuming 3.53 and 8 in **c**-axis and **ab**-plane directions, respectively. Then, the calculated conductivities for different θ are shown. The lower curve in Fig. 5-4c is the average of the theoretical predictions and it compares favorably with the experiment (dashed curve). The asymmetry in the curves arises from the small tunneling barriers compared to the large superconducting gaps.

Figure 5-5a shows a selected compilation of low-temperature results from tunneling measurements in different high-T_c materials, with different T_c values. The open circles are from tunneling through planar junction into **c**-axis-oriented films and thus should give $2\Delta_c$. For these data, $2\Delta_c/k_B T_c \approx 3.5$ is obtained in agreement with BCS. The other data, from point-contact measurements, yield $2\Delta_{ab}/k_B T_c \sim 5.4$, which is obtained from a least-squares fit to the data (excluding the film data that give Δ_c). This result may be taken as close to the Δ_{ab} gap.

Figure 5-5b shows the results for $\Delta_c(T)$ obtained from a planar junction for tunneling along the **c** axis. Although a reasonable fit to a BCS temperature dependence is obtained, T_c is reduced from what is found in the bulk. It is felt that this reduction may be due to the loss of oxygen near the Y123-metal boundary.

Some experimental results for the Tl compounds have been reported with gap-to-T_c ratios (Eq. 5-4a) $\sim$5. On the other hand, the energy gaps determined from $BaBiO_3$-type superconductors tend to yield results closer to 3.5, which is in good agreement with weak-coupled BCS.

Summarizing the tunneling measurements, it is probably fair to say that the gap-to-T_c ratio (Eq. 5-4a) gives the BCS result for the **c**-axis gap. However, a larger gap ratio is obtained in the **ab** plane, perhaps of the order of 4-6.

5-4c Specific Heat — As discussed in Section 2-5d, one of the parameter-free predictions of BCS is

$$\Delta C/\gamma T_c = 1.43 \qquad\qquad (5-4c)$$

where ΔC ($\equiv C_{es} - C_{en}$) is the electronic specific heat in the superconducting phase (C_{es}) minus that in the normal-metal phase (C_{en}), both evaluated at T_c and $\gamma T_c = C_{en}$ (Eq. 2-5e). This is shown in Fig. 2-5e, and Table 2-4 lists values for some conventional superconductors. This ratio is larger for strong-coupled superconductors (Table 2-4).

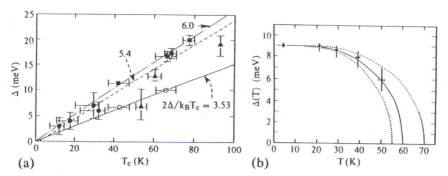

Fig. 5-5 (a) A selected compilation of superconducting energy gaps vs. T from La(n=1) and Y123 tunneling results. The lines corresponding to various $2\Delta/k_BT_c$ values are a guide to the eye. For details, see the article by Kirtley in the Notes. (b) $\Delta(T)$ from a planar junction for tunneling into a c-axis-oriented film of Y123. For details, see K. Hirata, K. Yamamoto, K. Iijima, J. Takada, T. Terashima, Y. Bando, and H. Mazaki, App. Phys. Lett. **56**, 683 (1990).

The problems faced, when comparable C vs. T results are sought in high-T_c materials, are illustrated in Fig. 5-6. First, since T_c is so large, the lattice-specific heat dominates the electronic contributions; the opposite regime occurs for most conventional superconductors (Fig. 2-7). Second, since H_{c2} is so large, it is not possible to measure C_{en} much below T_c, as can be done for most conventional superconductors. Third, in the measurements reported so far, at low temperatures usually a linear term and even an upturn in the specific heat (insert in Fig. 5-6) are found. These unexpected results are discussed.

As shown in an insert in Fig. 5-6, a specific heat jump (ΔC) for Y123 is observed in the vicinity of T_c. However, a wide range of ΔC values have been reported. After analyzing the shape of the C vs. T curve, a typical result is $\Delta C/T_c=55$ mJ/mole-K^2. To obtain the ratio in Eq. 5-4c, the value of γ must be known. No unambiguous method to obtain γ values in these materials is known, so Eq. 5-4c can be used to define a γ_{BCS} by

$$\gamma_{BCS} \equiv \Delta C/1.43T_c \qquad (5-4d)$$

and $\gamma_{BCS}\approx38$ mJ/mol-K^2 is obtained.

Obtaining reliable γ values has proved difficult. An approach has been taken by replacing some Cu by Zn, that is, $YBa_2(Cu_{3-x}Zn_x)O_7$, where T_c decreases with increasing x. From measurements of C vs. T for different Zn concentrations, ΔC vs. T_c can be determined. The results show

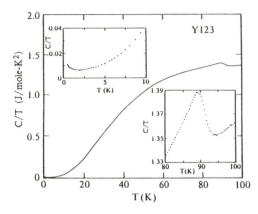

Fig. 5-6 The measured specific heat in Y123 plotted as C/T vs. T. The inserts emphasize the low-temperature data and the results near T_c. For details, see the Notes.

that ΔC scales with T_c and are in reasonable agreement with γ_{BCS} (Eq. 5-4d).

Another approach can be taken to determine γ. Estimates can be obtained from the Pauli contribution to the *magnetic susceptibility* χ_p,

$$\chi_p/\gamma = (\pi k_B/\mu_B)^2/3 \qquad\qquad (5-4e)$$

where μ_B is the Bohr magneton. This result (Eq. 5-4e) is sometimes called the **Wilson ratio**, which neglects spin fluctuations but seems to be a good approximation for heavy-electron materials (Section 2-8a) and other highly correlated systems. For Y123, magnetic susceptibility data can be fit to $\chi(T) = \chi_0 + C/(T - \Theta)$, with $\chi_0 \approx 2.8 \times 10^{-4}$ emu/mole. χ_p is obtained from χ_0 by subtracting estimates of the relatively small core, Landau, and van Vleck susceptibilities, which then, via Eq. 5-4e, yield what we may call γ_χ. Using these estimates, $\gamma_\chi = 35$ mJ/mole-K^2 is obtained, which compares well with γ_{BCS} as determined before.

These values of γ_{BCS} (or γ_χ) of about 38 mJ/mole-K^2 are much larger than found for typical metals (e.g., ~ 0.7 mJ/mole-K^2). Using γ to determine an effective mass (m^*), values of $m^* \approx 11$ are obtained. This effective mass is large, although much smaller than that found in the heavy-electron systems. It is interesting to note that the Wilson ratio (χ_p/γ) scales as well in the high-T_c systems as it does in the highly correlated heavy-electron systems; thus, this ratio is approximately independent of m^*. From our discussion of specific heat near T_c in Y123, it appears that the results may be in agreement with BCS (Eq. 2-5g). For the La(n=1) system with various dopings, the agreement between γ_{BCS} and γ_χ, for example, is not as good as described for Y123.

However, there is much less consensus than implied here about the correct γ value to use in Eq. 5-4c. Other values are obtained, for example, by extrapolating $d\gamma/dH$ found in Y123 powders to a fully superconducting sample; $\gamma = 16$ mJ/mol-K^2 is obtained (Notes). This is about half the value quoted above, and when used in Eq. 5-4c will give a jump in the specific heat about two times the (weak-coupled) BCS prediction, pointing to strong-coupled BCS.

To gain some experience in what might be expected from specific heat measurements in highly anisotropic superconductors, we mention results in the organic superconductor κ-(BEDT-TTF)$_2$Cu(NCS)$_2$ near $T_c \approx 9.4$ K (Notes). The specific heat, C, is measured without a magnetic field. Then, the sample is measured in a strong magnetic field of 5 T, C(H=5T), which is large enough (i.e., $H > H_{c2}$) to drive the sample normal, provided $T \gtrsim 3$ K. The $\Delta C = C - C(H=5T)$ jump at T_c is $\Delta C/T_c \approx 51$ mJ/mol-K^2. A γ value from measurements of the Pauli paramagnetism (Eq. 5-4e) is $\gamma = 34$ mJ/mol-K^2, from which $\Delta C/\gamma T_c = 1.50 \pm 0.15$ is quoted, in excellent agreement with weak-coupled BCS. However, in subsequent work, by using higher magnetic fields and lower temperatures, a direct determination of the normal-state electronic specific heat yields $\gamma = 25$ mJ/mol-K^2 or $\Delta C/\gamma T_c = 2.0$, which suggests strong-coupled BCS. Two points are clear from these results. First, even in extremely isotropic (conventional) superconductors, the $\Delta C/\gamma T_c$ ratio is similar to that calculated by BCS (Eq. 5-4c). Second, in this field, non-directly measured experimental data should be approached with caution. In the high-T_c field, this second point possibly extends even to some directly measured data because of sample preparation difficulties.

Low–temperature–results — A feature of the low-temperature specific heat in high-T_c superconductors is a linear term $\gamma(0)T$, and often an upturn (insert in Fig. 5-6). If this term is intrinsic, electronic states in the superconducting gap are implied, or the implication is the absence of a superconducting gap that is characteristic of conventional BCS superconductors. A low-temperature $\gamma(0)T$ term is consistent with the resonating-valence-bond theory and thus has attracted considerable attention. At present, whether this term is intrinsic or extrinsic has not been resolved. References in the Notes should be consulted.

In Y123 powders, which typically are used for the specific heat measurements, very small amounts of $BaCuO_2$ are usually found. This compound contains Cu^{2+} magnetic moments, which magnetically order near 10 K and produce a very large contribution to the low-temperature specific where $\gamma(0)$ is determined. It is possible that $BaCuO_2$ and small amounts of other impurities are the cause of $\gamma(0)$ or the upturn in C vs. T at low temperatures in Y123 (Fig. 5-6). At least this has been suggested by correlating $\gamma(0)$ with ΔC (at T_c) in the same Y123 samples.

In $(Ba_{0.6}K_{0.4})BiO_3$ superconductors, early measurements at low temperatures yielded nonzero $\gamma(0)$ values. Having a simpler chemistry, it appears that this material is easier to make pure even in the ceramic form, and recent measurements yield $\gamma(0) = 0$. Thus, at least for this superconductor, there do not appear to be any electronic states in the gap.

Summary — For most conventional superconductors, $\Delta C/\gamma T_c$ results are in agreement with BCS (Eq. 5-4c). Strong-coupled BCS give a ratio slightly higher than 1.43 and this usefully points out some of these materials (Section 2-6, Table 2-4). For high-T_c superconductors, the experimental results have, so far, been much less definitive, with many qualitative as well as quantitative problems remaining. However, calculations of $\Delta C/\gamma T_c$, and specific heat in general, for superconductors with larger electron-phonon coupling parameters (Sections 5-6f and 5-6i) and anisotropic gaps (Eq. 5-4b) may prove interesting.

Debye temperatures — By fitting measured specific heat (Fig. 5-6), Θ_D can be obtained. As always, a temperature-dependent Θ_D is obtained due to the limitations of the Debye model. In high-T_c crystals, the large number of optical-phonon branches undoubtedly make the Debye approximation even worse than for materials like Na or NaI, which have just one and two atoms in a primitive-unit cell, respectively. Thus, Θ_D should be expected to have a minimum $\sim\Theta_D/10$ or $\Theta_D/20$, of 10% to 20% (Notes). This is not important unless the Debye model is to be used seriously in order to obtain γT, or related quantities.

The Debye temperatures for the high-T_c materials lie in the 300 K to 450 K region; for example, for La(n=1) and Y123, respectively. For $(Ba_{0.6}K_{0.4})BiO_3$, one finds $\Theta_D=325$ K.

5-4d Infrared Results — The reflectivity of light from a metal is high. For a superconductor, for $\hbar\omega \leq 2\Delta$, the reflectivity should be 1. Then, above the superconducting energy gap, the reflectivity should decrease in a manner similar to the Drude reflectivity for a free-electron gas; at least, this is expected for conventional superconductors. Thus, $2\Delta(T)$ should be measurable by standard infrared (IR) techniques. However, the experiments are difficult because absolute reflectivities are needed and differences between R=1.00 and 0.99 can be important. For high-T_c crystals, the results are still controversial, and we briefly discuss some of the results.

The growth habit of Y123 yields crystals with much larger dimensions in the **ab** plane than along the **c** axis, making **ab** plane reflectivity measurements considerably easier. Such measurements from a single domain Y123 crystal ($T_c=90$ K) are shown in Figs. 5-7a and 5-7b for the IR electric field along the **a** and **b** axes. For $E \parallel a$, the reflectivity is approximately unity below 500 cm^{-1}. /he conductivities corresponding to the IR reflectivity results are shown in Figs. 5-7c and 5-7d. For $E \parallel a$ the conductivity is zero between 0 and 500 cm^{-1}. A similar threshold is observed for $E \parallel b$ except for a vertical displacement, which is interpreted as due to conductivity of the chains that are parallel to **b**. The results have been interpreted to yield a superconducting energy gap of 500 cm^{-1} ($2\Delta/k_BT_c \approx 8$) in the Cu-O planes and a chain conductivity that persists below 500 cm^{-1}.

Related IR measurements along **c** yield $2\Delta/k_BT_c \sim 3.5$. Thus, considerable anisotropy is suggested for the superconducting energy gap.

The normal-state conductivity of the Cu-O planes (Fig. 5-7c) falls with frequency less rapidly than would be expected form the free-electron Drude theory, with a fixed scattering rate, τ^{-1}. The data can be fit with a frequency-dependent scattering rate that is approximately proportional to frequency; $\tau^{-1} \approx 0.6\omega$ was found from the data in Fig. 5-7c. This frequency-dependent scattering rate is similar to what has been found in the PES measurements (Sections 4-7e and 5-4a).

However, other IR reflectivity as well as transmission experiments of Y123-oriented films have been interpreted rather differently. It is argued that these materials are in the clean limit, so the Drude peak is at very low energies (i.e., lower than 30 cm^{-1}). Thus, an opening of a superconducting gap in the Drude response will not be detectable. With this interpolation, it is suggested that features at $\sim$500 cm^{-1} are normal state (perhaps phonon) related, since they are observed above T_c.

In polycrystalline Y123 samples, the **ab**-plane superconductivity gap appears in a frequency region dominated **c**-axis phonons. However, for polycrystalline samples of $YBa(Cu_{1-x}Fe_x)_3O_y$ ($y \sim 7$), for $x \gtrsim 0.01$ the energy gap falls below the phonon frequencies and appears to be easily detectable. By measuring this feature as a function of temperature for different concentrations of Fe, $\Delta(T)/\Delta(0)$ can be measured, and the results are shown in Fig. 5-7e. As can be seen, there is good agreement with BCS theory. Unfortunately, from this data it is not possible to extrapolate back to x=0 to obtain the gap in pure Y123. In fact, even for the results shown in Fig. 5-7e, it is not clear whether the measured superconductivity gap is along the **c** axis or in the **ab** plane. On the other hand, it can be argued that results from polycrystalline samples will always accentuate the **ab**-plane gap

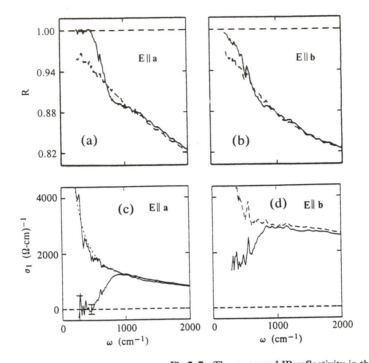

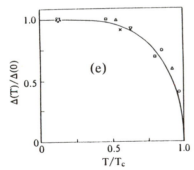

Fig 5-7 The measured IR reflectivity in the normal (T=100 K, dashed curves) and superconducting (T= 35K, solid lines) (a) for **E‖a** and (b) for **E‖b**. (c) and (d) are the corresponding conductivities in the normal and superconducting states. In (c), the dotted line is a fit to the normal-state conductivity with a frequency-dependent scattering rate. For details, see Z. Schlesinger, R. T. Collins, F. Holtzberg, C. Feild, S. H. Blanton, U. Welp, G. W. Crabtree, Y. Fang, and T. Z. Liu, Phys. Rev. Lett. **65**, 801 (1990). For other interpretations, see J. Orenstein et al., Phys. Rev. B **42**, 6342 (1990); F. Gao et al., Phys. Rev. B **43**, 10383 (1991). (e) $\Delta(T)/\Delta(0)$ vs. T/T_c for different Fe concentrations in $YBa_2(Cu_{1-x}Fe_x)O_y$ (y≈7). The symbols are: □, x = 0.2; o, x = 0.3; Δ, x = 0.04; ∇, x = 0.05; X, x = 0.06. The BCS prediction is the solid line. For details, see E. Seider, M. Bauer, L. Genzel., P. Wyder, A. Jansen, and C. Richter, Solid State Commun. **72**, 85 (1989).

because of the larger solid angle subtended by crystallites with this orientation. However, it is also possible that the Fe-atom substitution acts as pair breakers (Notes), reducing the effective anisotropy of superconducting gap as discussed in (Section 2-5e.)

5-4e Raman Results — Raman scattering in the conventional superconductors, Nb_3Sn and V_3Si, shows peaks below T_c, which are attributed to **electronic Raman scattering** by pairs of superconducting quasi-particles. The experimental results resemble the superconducting density of electron states (Fig. 2-6b) occurring close to $2\Delta(T)$. From these data, $2\Delta(0)$ values close to tunneling results are found and $2\Delta(T)$ vs. T is in agreement with BCS. In fact, by utilizing Raman selection rules, a gap anisotropy (Section 2-5e) of ~ 20 % is found in Nb_3Sn (Notes).

Natural, similar electronic Raman features sort after in high-T_c materials. Although an electronic scattering has been observed in Y123 as well as other high-T_c crystals, its origin is not clear. Unlike the structure observed in Nb_3Sn and V_3Si, in the high-T_c crystals, the intensity appears to increase linearly with frequency from $\omega \sim 0$ up to 400-600 cm^{-1}, depending on crystal and polarization. It appears as though there are electronic states in the gap. Also, using various Raman polarizabilities, this feature peaks at considerably different frequencies in the **ab** plane. More recent measurements have shown that these features do not scale with T_c and have other features that are not understood. Thus, the origin of this scattering is not yet clear (Notes) and the superconducting gap can not be obtained from these data, at least not with our present knowledge.

Using Raman scattering, a different approach has been taken to detect the superconducting gap. Various allowed, sharp phonon modes are observed in the high-T_c superconductors. The positions and line widths of these Raman active phonons can be studied as a function of temperature through T_c. If the phonons interact with superconducting gap, phonons with energies below $2\Delta(T)$ should be repelled to lower energy and those with energies above $2\Delta(T)$ should be repelled to higher energies. More sensitively, below T_c, phonons whose energies are above $2\Delta(T)$ should broaden because of a decrease in the relaxation time due to the increased density of electronic states above $2\Delta(T)$ (Fig. 2-6b).

Unfortunately, conflicting experimental results are found. All the materials studied have sharp T_c values at ≈ 92 K; however, they contain slight amounts of different impurities. One set of samples shows that the Raman line at 440 cm^{-1} increases in line width below T_c but the modes below ~ 300 cm^{-1} do not. The second set of samples shows that the 440 cm^{-1} line does not increase in line width below T_c but that the ~500 cm^{-1} does increase in line width below T_c. From the first set of samples, the gap is placed at $2\Delta/k_BT_c \sim 5.0$. However, from the second set of samples, the gap is placed at $2\Delta/k_BT_c \sim 7.0$. These differences suggest that the shape of the bands may differ in some important manner, or that the applicability of the theory to experiment may not be correct.

5-4f NMR Results — As discussed in Section 2-5f, a peak in the temperature dependence of certain properties is predicted just below T_c. This striking effect (Fig. 2-8) is called a **coherence peak**, and has been observed in measurements of the nuclear spin-lattice relaxation rate of conventional superconductors. The nuclear spin relaxation time (T_1) is the inverse of the relaxation rate, so $1/T_1$ should display a coherence peak. Indeed, for some of the conventional superconductors that have been measured, plots of $1/T_1$ vs. T peak just below T_c; then at lower temperatures the values decrease exponentially to zero. Such coherence peaks are considered to be a strong indication of the correctness of the BCS model for conventional superconductors.

Naturally, similar coherence peaks have been looked for in high-T_c materials. Nuclear-spin relaxation rates $(1/T_1)$ of copper and oxygen at all of the positions in Y123 have been measured. Namely, T_1 of Cu_p and Cu_c, as well as O_p, O_z, and O_c (Fig. 3-3) have been measured. However, coherence peaks have not been observed in any of these measurements. Rather, behavior similar to that of type I (Fig. 2-8) is found, and initially this has been taken as evidence against conventional BCS pairing.

It turns out, however, that conclusions from the lack of a coherence peak are less clear-cut than those from the observation of such a peak. Even in conventional superconductors, the magnitude of the observed coherence peak appears to be strongly effected by small anisotropies in the superconducting gap or finite lifetimes (strong-coupled BCS). This is shown by the alloying experiments and discussed in Section 2-6f. In high-T_c materials, the observed $1/T_1$ data can be explained by invoking any sort of temperature-dependent pair-breaking interaction (Notes), such as spin fluctuations from conduction electrons. Large anisotropies in the superconducting gap probably occur and spatial inhomogenities also probably reduce a coherence peak in the high-T_c materials. Theoretical calculations of strong-coupled BCS have been shown to reduce the magnitude of the coherence peak (Notes). Thus, it appears to be difficult to draw conclusions from the lack of observation of a coherence peak in the high-T_c materials.

5-5 Isotope Effect

The isotope effect (1950) first showed that the atomic mass and, therefore, the electron-phonon interaction play an important role in superconductivity. The experimental results are described in terms of

$$T_c \propto M^{-\alpha}$$
$$\alpha = - d(\ln T_c)/d(\ln M) \qquad (5 - 5a)$$

where M is the atomic mass and α is a parameter. The first experiments were on the superconductor mercury (Hg) and gave $\alpha = 0.50 \pm 0.03$. It was fortuitous that Hg was first studied, since other conventional super-conductors yield $\alpha \neq \frac{1}{2}$. Nevertheless, the isotope effect pointed the way to the electron-phonon interaction, which resulted in an understanding of electron pairing.

The weak-coupling BCS theory predicts that $T_c \propto \omega_q$, where ω_q is a typical frequency of the phonons that are important in the electron-phonon interactions. The BCS result is

$$k_B T_c = 1.14 \, \hbar \omega_q \, e^{- 1/N(E_F)U_0} \qquad (5 - 5b)$$

where $N(E_F)$ is the density of states at the Fermi energy. U_0 is the electron-electron attractive strength that reflects the competition between electron-electron Coulomb repulsion forces and the phonon-mediated electron-electron attractive forces. The isotope effect is built into BCS be-cause they postulated a cutoff in the attractive electron-electron interaction at the Debye energy $k_B \Theta_D$, which then replaces $\hbar \omega_q$ in Eq. 5-5b. From el-ementary theory, the Debye temperature $\Theta_D \propto M^{-\frac{1}{2}}$ so Eq. 5-5a, with $\alpha = \frac{1}{2}$, follows. Thus, BCS predicts $\alpha = \frac{1}{2}$ for all weak-coupled supercon-ductors.

In Table 5-1, experimental α values are listed for some elemental superconductors, and it is clear that a variety of results have been found. First, it can be seen $\alpha \approx 0.5$ only for the s-p metals; that is, only for metals whose valence band is composed principally of s and p electrons. Second, for many transition-metal superconductors (Ru, Os, Mo, Zr) and their compounds (Nb_3Sn), α values are in the 0.0 to 0.3 range. It has been argued that the small isotope effect is associated with the fact that the $d^5 s^1$ elec-tronic configuration has a high lattice stability, and is only a few eV above the ground-state, half-filled d^6 electron configuration. Third, it has been suggested that superconductivity in α-U (and La) is due to an attractive interaction associated with a virtual f-electron exchange and does not result from a phonon-mediated pairing mechanism. The other results in Table 5-1

Table 5-1 Experimental isotope values (α_{exp}) for some elemental superconductors along with calculated values. See the article by Meservey and Schwartz in Parks, Ed. (Bib.) for references.

Element	α_{exp}	α_{SWI}	α_{MA}	α_{G1}	α_{G2}
Zn	0.45±0.01	0.2	0.35	0.40	0.415
Cd	0.50±0.10	0.2	0.34	0.37	0.385
Hg	0.50±0.03	0.4	0.46	0.465	0.48
Tl	0.50±0.10	0.3	0.43	0.45	0.48
Sn	0.47±0.02	0.3	0.42	0.44	0.455
Pb	0.48±0.01	0.3	0.47	0.47	0.485
Zr	0.0		0.30	0.35	0.15
Mo	0.33±0.05	0.15	0.3	0.35	0.35
Ru	0.0 ±0.10	0.0	0.35	0.0	0.065
Os	0.20±0.05	0.1	0.25	0.1	0.225
Re	0.39±0.01		0.41	0.3	0.355
U(α)	−2.2±0.2	-	-	-	-

are from theoretical calculations and the original references should be consulted for details. Besides uranium, a reverse-isotope anomaly has been found in palladium hydride (PdH); $T_c \approx 9$ K for the hydride, but the deuteride (PdD) has a *higher* transition temperature ($T_c \approx 11$ K).

Most of the elemental superconductors have one atom per primitive unit cell and thus only acoustic phonon branches, implying that the Debye model might be appropriate. However, the high-T_c superconductors are chemically complicated, with many atoms in a primitive unit cell (Chapter 3) leading to optical branches at low as well as high energies. Thus, more complicated results may be expected.

Isotope measurements in the high-T_c superconductors are mostly performed by replacing ^{16}O with ^{18}O, because it is thought that O-atom vibrations (the highest-frequency phonons) might be responsible for the major part of the electron-phonon interaction. Also, oxygen replacement gives the largest fractional change of mass. The isotope mass exponent α (Eq. 5-5a) implicitly contains contributions to the isotope shift in T_c from all of the constituit atoms. Thus,

$$\alpha = \Sigma_i \alpha_i \qquad (5-5c)$$

might be expected where α_i is the partial isotope exponent for element i. However, the measured change of T_c with isotope substitution of atoms other than oxygen atoms is very small. Also, one could take the point of view that the replacement of Y in Y123 by rare earths (RE123) might be thought of as an isotope substitution. This replacement then would be equivalent to an extremely large mass change (Eq. 5-5a); however T_c does not change. Thus, to a first approximation, α_i from the other atoms are ignored, and only results from oxygen isotope measurements are discussed, using α without a subscript.

For oxygen replacement in Y123 with $T_c \approx 92$ K, very small values of α have been found (between 0.0 and 0.1). Similarly, small values of α have been reported for other high-T_c materials with $T_c > 77$ K, such as 2-Bi(n=2 and 3), where shifts of $T_c \approx 0.3$ K are found. However, caution must be exercised because of the experimental difficulties. It is difficult to guarantee the same carrier concentrations in the ^{16}O and ^{18}O samples and T_c is very sensitive to these concentrations.

Figure 5-8 shows the complex α vs. x behavior found in $(La_{2-x}Sr_x)CuO_4$. The highest T_c material (x=0.15) has a small α. However, material with slightly lower x still yield large T_c values. However, these materials have α values in the range of 0.5 and even greater than 0.5. Thus, although the highest-T_c materials have small α, the situation is complex. If α vs. x for $(La_{2-x}Ba_x)CuO_4$ is plotted in Fig. 5-8, essentially the same curve is found, except that this material appears to not be superconducting in the x$\approx$0.12 region. This lack of superconductivity may be due to a structural phase transition at higher temperatures (Section 3-3). Nevertheless, in the orthorhombic phase, both the Sr- and Ba-doped La(n=1) have the same unusual α vs. x curves, see Fig. 5-8 and the references in the caption. Since the masses of Sr and Ba are so different, it might be argued that the anomalous α vs. x behavior is due to electronic effects, possibly associated with band filling effects or peaks in the electric density of states (Section 5-6h).

Interestingly, very recent results of α vs. x in $(Y_{1-x}Pr_x)Ba_2Cu_3O_7$ and $Y(Ba_{2-x}La_x)Cu_3O_7$ show that as T_c decreases with increasing x, α increases to $\alpha \sim 0.4$ for $T_c \sim 40$ K. Isotope effect measurements are always difficult in these cuprate oxides. Hence, we await other results in Y123-based as well as other high-T_c superconductors, where T_c can be varied by changing the doping. $(Ba,K)(Pb,Bi)O_3$ superconductors exhibit α values in the 0.5 range.

Summarizing, we see that a wide range of α values has been found in high-T_c materials. For materials with T_c in the 92 K range, very small values have been reported. For La(n=1), complicated α vs. x values are found, as shown in Fig. 5-8, including ones larger than the BCS value of ½; this is quite unusual. It is probably fair to say that the value of α from the isotope

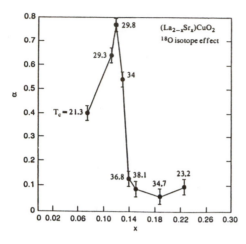

Fig. 5-8 The isotope shift α vs. x for ^{18}O replacing ^{16}O in $(La_{2-x}Sr_x)CuO_4$. The numbers next to each data point are the T_c values. For details, see M. K. Crawford, M. N. Kunchur, W. E, Farneth, E. M. McCarron, III, and S. P. Poon, Rev. B **41**, 282 (1990) and Science **250**, 1390 (1990).

effect, particularly in narrow-band metals that include the high-T_c materials, neither proves nor disproves phonon-mediated superconductivity. However, it is probably also fair to say that finite α values indicate that phonons are involved in the pairing mechanism to some extent. Some aspects of this are discussed in Section 5-6h.

5-6 The Pairing Mechanism

5-6a Introduction — The nature of the pairing mechanism in high-T_c crystals is not understood at present. Its mechanism seems to be completely "up in the air," with different camps totally devoted to their own favorite theory. At first, pairing mechanisms of magnetic origin were in favor because, in their insulating phases, the high-T_c materials are typically in an antiferromagnetic state (Figs. 4-4c and 4-5b). Also, the magnetic exchange energies (J_{exc}) are ~ 4 times larger than the phonon energies. Since $T_c \propto \Theta_D$ for phonon-mediated pairing (Eqs. 2-6a and 5-5b), it would seem that some sort of magnetic-mediated pairing, where $T_c \propto J_{exc}$, would yield higher T_c values, as found in the high-T_c materials. The unusual RVB model of these materials has been, and continues to be, discussed. Some of the motivation for the RVB model arises from the strange normal state properties of these materials (Section 4-7b).

However, by 1990, phonon-mediated pairing in these high-temperature superconductors was being taken seriously. (The reader should note the discussion on prejudgements in the Preface.) Certainly, phonon-mediated pairing is consistent with the experimentally observed s-state pairing (Section 5-2).

Besides the reasons mentioned above, there were many other reasons that phonon-mediated superconductivity in the cuprates as, at first, thought unlikely. First, T_c values were too high. Not only was $(T_c/\Theta_D)^2$ not small (Tables 2-2 and 2-4), but $T_c \sim \Theta_D$. Thus, extremely strong-coupled BCS was thought to be required, and such a large electron-phonon parameter (λ_{ep}) would likely cause a structural phase transition (Section 3-3).

Second, the initial isotope experiments gave α values close to zero (Section 5-5), which seemed to eliminate phonons as a possibility. However, transition metals also can have α values approaching zero (Table 5-1). Thus, $\alpha \neq \frac{1}{2}$ neither proves nor disproves phonon-mediated superconductivity, as discussed at the end of Section 5-5.

Third, specific heat measurements gave a linear term in the specific heat below T_c (a nonzero γ, Section 5-4c), which were taken to imply that allowed states were in the superconducting "gap," or even that there was no gap, in the BCS sense. Other experiments, such as electronic Raman measurements, also pointed to states in the gap.

Fourth, the observed normal-state linear resistivity (Section 4-3) was taken to mean that these layered metallic cuprates were different kinds of metals than those usually encountered. The possibility was suggested that they were not Fermi-liquids (i.e., they did not have conventional E vs. k curves), but were more exotic metals; they might then be described by the resonating-valence-bond theory (Section 4-7b).

Fifth, the initial reports of gap-to-T_c ratio seemed to vary from about 2 to 15, so it was felt that something was amiss. However, the experimental results have settled down. Nevertheless, the results are still greater than 3.52 (Eq. 5-4a).

However, experimental evidence is beginning to be published that points more and more to phonon-mediated pairing, perhaps with some additional booster. The isotope-shift data in Fig. 5-8 are an indication of the importance of phonons; they appear to give no support to magnetic or excitonic pairing. Also, the isotopic shift data in Y123-related materials with lower T_c values, discussed at the end of Section 5-5, is another indication of the importance of phonons.

In this section, other support for phonon-mediated pairing will be given. There is now more theoretical understanding of how it would be possible to obtain such high values of λ_{ep} and hence T_c with phonons without the crystal becoming unstable with respect to a phonon-induced structural phase transition. Also, more theoretical understanding has tended to eliminate some of the objections discussed above.

Thus, this entire section is written, and the topics are chosen, with a prejudgement toward the phonon-mediated pairing camp. A more straightforward warning to the reader can not be issued — be aware.

5-6b Soft Phonon Modes — Using the Eliashberg formalism for BCS phonon-mediated pairing and a very strong electron-phonon   parameter (λ_{ep}, Section 2-6), it has been suggested that T_c values as high as 40 K could be obtained. For higher T_c's, nonphysical parameters are thought to be required. However, an order of magnitude enhancement has been calculated in λ_{ep}, if it is assumed that some atoms (the oxygen atoms, for example) are in a double-well potential.

La(n=1) — The evidence for a double-well potential in La(n=1) is clear. The phase transition between the high-temperature tetragonal and low-temperature orthorhombic phase (Fig. 3-7 and Section 4-6) is due to a soft, Brillouin-zone-edge phonon. That is, in the high-temperature phase, as temperature decreases, the energy of a phonon with a **k** value at the Brillouin-zone boundary decreases, or becomes "soft." Finally, its eigenvector "freezes in ," yielding a low-temperature phase with static displacements closely resembling the eigenvector of the phonon in the high-temperature phase. Then the single-atom potential in the high-temperature phase can be described in terms of a double-well potential. However, for most x values, the structural phase transition in La(n=1) occurs at temperatures far from T_c and, at least in the Sr-doped material, it appears to not be strongly correlated with T_c even when the two temperatures are close. In Ba-doped La(n=1), a first-order transition to a low-temperature tetragonal phase (Section 3-3) seems to hinder superconductivity. Thus, in La(n=1), firm evidence for T_c-enhancement being correlated with a structural phase transition appears to be tenuous.

Y123 — In the 93 K superconductor $YBa_2Cu_3O_7$, there is no evidence for a structural phase transition. Early Raman and infrared (IR) measurements showed that some of the modes in the $\omega \sim 300$ cm^{-1} region have discontinuities in $d\omega/dT$ at T_c. However, the changes in ω for these modes between 0 K, T_c, and 300 K is only 1% or 2%, with hardly any changes in line shape. Thus, these observed modes do not appear to arise from double-well or even anharmonic potentials.

Recent extended x-ray absorption fine structure (EXAFS) measurements on Y123 have been refined to yield a best fit to the data with the apical oxygen (O_z, Fig. 3-1) in a double-well potential, with the distance between the two wells varying with temperature from 0.13Å at 10 K to 0.02Å near T_c. These are interesting results, if correct. However, very careful, more standard diffraction approaches have been taken. These include a joint x-ray and neutron diffraction refinement of the Y123 structure and a pulsed-neutron powder diffraction refinement of the Y123 structure (Notes). These studies fail to show anything unusual happening to any of the atom positions; in particular, the temperature dependence of the Cu_c-O_z or Cu_p-O_z distances and the anisotropic thermal factors for O_z seem well behaved between 10 K and room temperature. More work remains to be done, but firm evidence for unusual behavior near T_c of O_z or the other oxygen atoms appears to be problematic (Notes).

5-6c Temperature-Dependent Phonon Modes — In Section 5-6b, soft modes in these materials are discussed; that is, phonon modes whose frequency approaches zero as the temperature is lowered from some high temperature. In this section, examples of much milder temperature-dependent phonon effects are considered; that is, phonons whose frequency shifts ($\Delta\omega$) with temperature amount to a few percent for $\Delta\omega/\omega$. These observations are usually determined by Raman scattering. Also discussed are some larger effects measured by ion channeling in combination with Rutherford back scattering spectrometry (RBS) and related techniques.

Phonon shifts amounting to $\Delta\omega/\omega \sim 2\%$ between T_c and 0 K have so far been observed in Y123 and 2-Bi(n=2). The atoms principally involved in the phonon motion are the planar oxygens O_p and the apical oxygens O_z (Figs. 3-1 and 3-2), respectively. These small shifts could be due to gap effects, as discussed in Section 5-4e, or could be associated with slight atom shifts at T_c. The atomic motion involved for these small phonon shifts, in fact most of the phonons discussed in Section 5-4e, is motion along the c axis, because only such phonons experimentally are found to have significant Raman intensity. Thus, there is little Raman data for in-plane atom motion.

Since all of the high-T_c structures have a center of symmetry (Chapter 3), the Raman and infrared (IR) modes are mutually exclusive (the **exclusion rule**). Hence, IR measures different phonons than Raman. However, for IR active phonons, the large conductivity in the planes screen the long-range forces, reducing the LO-TO splitting to ≈ 0, which again makes IR phonons with in-plane atom motion not observable. IR active phonons with atom motion along the c axis are screened less and some small frequency shifts have been observed.

As a result of intensity effects in both Raman and IR measurements, phonons with atom motions in the **ab** plane have hardly been measured by these techniques. However, such motion can be studied by ion channeling in combination with Rutherford backscattering (RBS), particle-induced x-ray emission, resonant neutron absorption spectroscopy, and other techniques where motion of selective atoms may be sensed but without energy or **k**-vector selectivity.

In **ion channeling** experiments, an ion beam (e.g., 1.5 MeV ^{4}He$^+$) is incident on a single crystal along a major crystallographic direction (e.g., the [001], or **c** axis, in Y123). At the exact angle along this direction, the ions are steered between the atomic planes by a series of correlated small-angle collisions. This channeling effect severely reduces the RBS yield. The critical angle of incidence, below which the channeling occurs, can be directly related to the atomic thermal vibrational amplitudes. By energy resolving the RBS ions, different atoms in the crystal can be sampled. For example, heavier atoms provide larger recoil momentum to the channeling ions. In this manner, the channeling angle full-width at half-maximum (FWHM) is measured vs. temperature. In both Y123 and 2-Bi(n=2), with decreasing temperature, the channeling angle FWHM abruptly decreases below T_c, indicating a decrease in the Cu-atom displacements perpendicular to the **c** axis.

These channeling measurements have been further corroborated by **neutron resonance absorption spectroscopy** (NRAS) data in 2-Bi(n=2), where results from different constituent atoms in the crystal can also be selected (Notes). Again, a sharp decrease of the in-plane Cu-atom motion is observed below T_c. The NRAS data also show no change of the **c**-axis Cu-atom motion.

Thus, the ion channeling and related experiments appear to point to sharp phonon anomalies at T_c. The phonons involve principally Cu (and probably O) atom motion in the **ab** plane. As discussed, this type of motion is difficult to observe by Raman or IR techniques. However, neutron diffraction should be able to detect these vibrations and provide energy and wavevector resolution so that specific phonons can be studied. Clearly, these types of results are of great interest (Notes).

5-6d Neutron Measurements — Inelastic neutron scattering from single crystals yields energy as well as **k** resolution. Thus, phonon dispersion curves, E vs. **k**, can be measured. This is typically done along a few important **k** directions. Measurements in high-T_c crystals should be rather informative in light of the ion channeling results discussed above. The problem is that relatively large single crystals (~ 1 cm^3) are required for in-elastic neutron measurements, especially since most high-T_c materials have many atoms in a primitive cell and hence many phonon branches. (For Y123-O$_7$, there are 3×13.) For most high-T_c materials, large, high-quality crystals are still difficult to obtain. Lacking large single crystals, the going **generalized phonon density of states**, $G(\omega)$, can be measured from a sintered powder of the material. $G(\omega)$ differs from the **phonon density of states** $F(\omega)$ by weighting the vibrations of the ith atom with the ratio σ_i/M_i. (σ_i is the bound scattering cross section and M_i is the atomic mass of the ith atom.) This ratio is about the same for La, Y, Sr, and Cu; however, for O it is increased by about 2.5. Thus, where O-atom vibrations are important, G is enhanced compared to F.

The measured generalized phonon density of states for the super-conductor Y123-O$_7$ and the insulator Y123-O$_6$ are shown in Fig. 5-9a. In going from the insulator to the superconductor, there is a general softening (decrease in energy) of the high-energy phonon modes, and a hardening (increase in energy) of the low-energy modes. The hardening of the low-energy spectra is in agreement with the stiffening of the transverse acoustic branches in the superconductor. A great deal has been said about this soft-ening and how it fits with what might be expected for phonon-mediated pairing (Section 2-7). Preliminary single-crystal E vs. k results from Y123 are in agreement with the density of states data (Fig. 5-9a).

However, it has been suggested that softening of the high-energy phonon modes in going from O$_6$ to O$_7$ actually has little to do with a general weakening of the phonon force constants. In Y123-O$_6$, the chain copper atoms (Cu$_c$) have two-fold coordination with the apical oxygen atoms (O$_z$), which form O$_z$-Cu$_c$-O$_z$ "sticks" (Section 3-2f). In Y123-O$_7$, Cu$_c$ at-oms are four-fold coordinated to 2O$_z$ plus 2O$_c$(b) (Fig. 3-3). The highest-frequencey modes at $k \sim 0$ are Raman and infrared phonons with motion along the c axis; for the sticks in Y123-O$_6$, the frequencies are 600 cm^{-1} and 650 cm^{-1}, respectively. The corresponding vibrations in Y123-O$_7$ oc-cur at 505 cm^{-1} and 575 cm^{-1}. Thus, it would appear that the decrease in frequency of the highest-frequency branches observed in the phonon den-sity of states, in going from O$_6$ to O$_7$ material, has little to do with a general softening of vibrational force constants; it is due to a coordination change of the chain-Cu atoms.

The phonon density of states for 2-Bi($n=2$) is shown in Fig. 5-9b at three temperatures: 5 K, just above T_c, and room temperature. The changes in $G(\omega)$ are so small that there is little need to distinguish between the temperatures. Related measurements (not shown) have been made in the

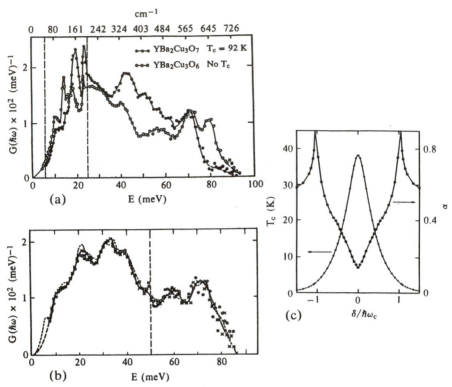

Fig. 5-9 (a) Phonon density of states, measured at 6 K, for Y123-O$_7$ and Y123-O$_6$ as indicated. (b) Phonon density of states for 2-Bi(n=2) with T$_c \approx$80 K. The measurements were made at 5 K ($\times$), 120 K (o), and the dashes correspond to higher-resolution data at 296 K. For details, see B. Renker, F. Gompf, E. Gering, D. Ewert, H. Rietschel, and A. Dianoux, Z. Phys. B **73**, 309 (1988), and ibid **77**, 65 (1989). (c) Calculated T$_c$ and α as a function of the Fermi-energy shift (δ) normalized to a phonon cutoff frequency ($\hbar\omega_{co}$). For details, see C. C. Tsuei, D. M. Newns, C. C. Chi, and P. C. Pattnaik, Phys. Rev. Lett. **65**, 2724 (1990). References to related work can be found in this paper and another recent paper: R. S. Markiewicz, Physica C **169**, 63 (1990).

insulator phase by replacing Y with Ca. In this case, the changes in G(ω) also are relatively small, certainly smaller than observed in Y123 (Fig. 5-9a) and thus in agreement with the proposal that the softening effect in Y123 is due to a change of the chain-Cu atom coordination and not due to a general softening of vibrational force constants.

It appears that measurements of G(ω) vs. T for a given superconductor show negligible changes at T$_C$. Yet ion-channeling results indicate the phonons involving Cu motion in the **ab** plane should change at T_c. We await complete phonon dispersion curves to learn exactly what these Cu-atom motions involve.

La_2CuO_4, and its insulating isomorph La_2NiO_4, can be more easily grown in larger crystals than Y123. Also, they have only $3 \times 7 = 21$ phonon branches, thus are easier to measure. Interesting anomalies are observed in the high energy, in-plane Ni-O stretching vibrations. However, the corresponding Cu-O vibrations in La(n=1) do not show these anomalies. Thus, the meaning of these effects, and lack of effects, is not as yet clear. References in the Notes should be consulted.

5-6e High–Energy Tunneling Results — As mentioned in Section 2-7c, a phonon-like structure on the single-particle tunneling current demonstrates the importance of the phonon-mediated pairing mechanism in conventional superconductors. In fact, whatever boson-mediated pairing mechanism is important for the superconductivity, its signature should appear in tunneling measurements. Thus, one asks: what are the tunneling results in the high-T_c superconductors? The answer is that they are difficult to carry out because the voltages are rather large due to the large gaps and the junctions tend to have high leakage currents.

Tunneling conductance has been reported from 2-Bi(n=2)-GaAs junctions at 4.2 K. The results indicate a superconductivity gap with $2\Delta(0)/k_BT_c \approx 6.0$ and a great deal of structure is observed above the gap. The structure is in reasonable agreement with the phonon spectra of 2-Bi(n=2). Thus, it appears that phonon-mediated pairing is important in this material, although the data do not exclude continuations from other mechanisms (Notes).

In recent related measurements, point-contact tunneling in the electron-doped superconductor, $(Nd_{2-x}Ce_x)CuO_4$ with $T_c=22$ K, have yielded sharp conduction peaks at $V=\pm\Delta/e$ with low leakage currents. Furthermore, for these junctions symmetric reproducible structures are observed at high-bias tunneling conductances. By inverting the tunneling data (Section 2-7c), $\alpha^2F(\omega)$ vs. ω is obtained. It is premature to discuss these results in detail because of measurement uncertainties and $F(\omega)$ is not presently known for Nd(n=1). However, the peaks in $\alpha^2F(\omega)$ have similarities to the measured IR and Raman $k\approx 0$ frequencies (Notes). These tunneling data yield $2\Delta/k_BT_c = 3.9$ for Nd(n=1) at low temperatures, and from the inversion analysis $\lambda_{ep} = 1.0$ and $\mu^* \approx 0.08$ are obtained.

In summary, tunneling measurements in 2-Bi(n=2) and Nd(n=1), which are hole and electron conductors, respectively, indicate that phonon-mediated pairing is important in these superconductors. It is not apparent to what extent this excludes other contributions to the pairing energy. However, these results appear to point toward phonon-mediated pairing and strong-coupled superconductivity.

5-6f Electron–Phonon Coupling Parameter Calculations — Calculations of the electron-phonon coupling parameter, λ_{ep}, are discussed in Section 2-6d. There it was noted that large λ_{ep} could be obtained for high-frequency phonons in spite of some belief to the contrary. Naturally, such calculations have been considered in high-T_c materials. However, as can be appreciated from Section 2-7d, the calculations are involved and computer-intensive. It may take time to theoretically sort out the phonons that couple more strongly to the electrons if, indeed, some particular phonons can be singled out.

There seems to be an important difference between conventional and high-T_c superconductors. In fairly good metals, which include most of the conventional superconductors, the conduction electrons screen out, in very short distances, changes in the potential due to core displacements. However, this is not the case for the high-T_c metals, which have a stronger ionic character. For example, in the insulator phases, the longitudinal-transverse optic splittings large, with only a small background screening constant, $\varepsilon_\infty \sim 4$-$6$; even in the metallic phases, the low density of charge carriers is largely confined to the Cu-O planes. As a result, ion-core motion perpendicular to the planes is hardly screened, and motion in the planes is poorly screened. Thus, even distant ion-core displacements may change the local potential at a charge carrier site. These non-local ionic contributions to the electron-phonon coupling parameter can be large and appear to yield larger than expected $\lambda_{ep}(q\nu)$ values for higher-energy phonons. As discussed in Sections 2-6d and 5-6i, high values of T_c may result.

As discussed in Section 2-6d, λ_{ep} must be integrated over some representative part of the Brillouin zone and calculations away from the zone center are presently difficult. However, theoretical estimates are now appearing with λ_{ep} values of between 1 and 2.4 (Notes). These are extremely large for a material that is stable (no soft modes, Section 5-6b). Further, preliminary experiments also seem to yield $\lambda_{ep} \sim 1$ (Notes).

5-6g Electron–Phonon Coupling Parameter Measurements — Few direct measurements of the electron-phonon parameter, λ_{ep} (Eq. 2-7a), have been devised. Theoretical estimates of it are possible (Sections 2-7d and 5-6f), or interpretation of experiments such as resistivity (Section 4-3).

However, recent femtosecond (1 fs $= 10^{-15}$ sec) pulse-probe measurements of the electron temperature can be interpreted to yield values of λ_{ep}.

In these pulse-probe experiments, a femtosecond duration pulse of light from a mode-locked laser impinges on the metal sample (which can be at room temperature). The light is first absorbed by the electron system, which heats and is thus out of thermal equilibrium with the atoms in the crystal. The rate of temperature equilibration between the electron and atom temperature is governed by λ_{ep}. The electron temperature approaches that of the atoms (T_e and T_a, respectively) as given by a relaxation rate equation,

$$T_e \frac{\partial T_e}{\partial t} = - \frac{3\hbar}{\pi k_B} \lambda_{ep} < \omega^2 > (T_e - T_a) \qquad (5-6a)$$

where $< \omega^2 >$, the second moment of the phonon spectrum, can be calculated or related to the Debye frequency.

From the experimental point of view, the key is that changes in T_e must cause changes in the reflectivity. It is these changes of reflectivity, measured by the probe beam as a function of delay time after the pump beam, that are measured.

Using this femtosecond, pulse-probe technique, many ordinary metals as well as superconductors have been measured. Small values such as $\lambda \approx 0.1$ have been obtained for Cu and Ag, which are not superconductors. Values of $\lambda_{ep} \approx 1.16$, 1.45, and 0.83 are obtained for the strong-coupled superconductors, Nb, Pb, and V_3Ga, respectively. For the high-T_c superconductors, Y123 and 2-Bi(n=2), the measurements yield $\lambda_{ep} \approx 0.90$ and 0.82, respectively. Furthermore, for the insulator Y123-O_6, $\lambda_{ep} \approx 0.07$, which is in sharp contrast to the strong-coupled value found for the metal.

There are many questions that can be raised concerning these experiments. Certainly, one of them is, how should $< \omega^2 >$ be estimated for the high-T_c materials where there are many low- and high-frequency phonons. Do all of the phonons get averaged or only the ones that are important for the pairing interaction? And which ones would those be?

5-6h Phonons Plus Electron Density of States Singularity — Even in the conventional A15 superconductors, T_c enhancements due to sharp singularities in the electronic density of states at, or very near, the Fermi surface have been proposed. For high-T_c crystals, such a singularity could be associated with the density of states, N(E), that arises from an approximately two-dimensional Fermi surface (Section 4-7d). It has been felt that disorder due to doping and structural defects would tend to smear any sharp peak of $N(E)$ near E_F. Furthermore, the effectiveness of, for example, a van Hove singularity for a T_c-enhancement should greatly diminish as E_F shifts from the singularity (Section 4-7e).

However, recent slave-boson, mean-field band structure calculations for the cuprates indicate the E_F can be pinned close to a logarithmic van Hove singularity. We discuss a few of these results because they yield high-T_c values with modest electron-phonon parameters (weak-coupling) and also can be used to calculate the isotope shifts (α, Section 5-5). See the Notes for references.

A logarithmic density of electronic states is assumed, such as

$$N(E) = N_0 \ln |E_F/(E - E_F)| \qquad (5 - 6b)$$

where N_0 is the density of states normalized to a flat band of width $2E_F$. Then, using this $N(E)$, T_c and α can be calculated from the standard BCS gap equation. To do this, the pairing energy is integrated over $E_F \pm \hbar\omega_{co}$, where ω_{co} is a phonon cutoff frequency that probably may be taken as the maximum phonon frequency $\sim \omega_{max}$ (Eq. 2-7a), so a cutoff temperature can be given as $\hbar\omega_{co} = k_B T_{co}$. Also, if the shift of E_F from the singularity is taken as δ, where it is assumed that $\delta < 2k_B T_c$, then solving the gap equation to obtain T_c and calculating $\alpha = - \partial \ln T_c/\partial \ln M$ (Eq. 5-5a), it is found that

$$T_c \approx 1.36 T_F \exp\left[-\left\{ \frac{2}{N_0 U_0} + \left(\ln \frac{T_F}{T_{co}} \right)^2 - 1 + \frac{\delta^2}{2k_B^2}\left(\frac{1}{4T_c^2} + \frac{1}{T_{co}^2} \right) \right\}^{1/2} \right]$$

$$\alpha \approx \frac{1}{2}\left[\left\{ \ln\left(\frac{T_F}{T_{co}} \right) + \frac{\delta^2}{2k_B^2 T_{co}^2} \right\} \left\{ \ln\left(\frac{1.36 T_F}{T_c} \right) - \frac{\delta^2}{8k_B^2 T_c^2} \right\}^{-1} \right] \qquad (5 - 6c)$$

where U_0 is the BCS electron-phonon matrix element (Eq. 2-6a). These equations show several important points. First, the prefactor in the T_c equation is now an electronic temperature, T_F, rather than a phonon temperature, Θ_D, and T_F can be larger than Θ_D. Second, T_c can be high, peaking when E_F is just at the singularity (Fig. 5-9c). Third, when T_c peaks, α is a minimum, ~ 0.1. Fourth, for low-T_c values obtained by varying the doping, α can increase to above 0.5 (Fig. 5-9c). These last two points are in qualitative agreement with the measured isotope effect shown in Fig. 5-8.

Thus, it appears that within the phonon-mediated electron pairing BCS framework, with a logarithmic van Hove singularity near E_F (or probably any strong energy-dependence in the density of states), high-T_c values can be accounted for. Further, the α vs. doping is calculated (Fig. 5-9c) that has qualitative and quantitative features in agreement with measurements (Fig. 5-8).

The T_c vs. doping result in Fig. 5-9c was computed for $\lambda_{ep} = 0.25$ (weak-coupling BCS); T_c peaks at ~40 K. For a $\lambda_{ep} = 0.36$ (still weak-coupling BCS), T_c peaks at ~92 K. Thus, even weak-coupling with a singularity in the election density of near E_F is a potent way to increase T_c within the framework of phonon-mediated superconductivity.

However, it is too early to determine the generality of these ideas for high-T_c superconductors. Experiments to test some of these ideas immediately come to mind. PES (Sections 4-9 and 5-4a) should be able to measure a sharp peak in the density of electronic states; perhaps the relatively flat bands near $\overline{M}$ (Fig. 4-9a) may be related to the required peak. The shift of this peak with doping could also be looked for with PES. Measurements of α vs. doping in other high-T_c systems would be of distinct interest, especially if the doping could be varied on both sides of the T_c peak. An understanding of the relation between a sharp peak in the electron density of states and normal state properties might also be enlightening.

By using this model, involving phonons plus a van Hove singularity in the electron density of states, several other interesting results have been calculated. A T_c vs. doping having about the correct width (Fig. 4-5c) has been calculated. Also, the small correlation widths in the **ab** plane can be accounted for.

5-6i Phonons Alone — In very recent work (Notes), numerical solutions of the Eliashberg equations for large electron-phonon parameters (λ_{ep}) were reinvestigated. The results indicate that for large λ_{ep}, higher T_c values are obtained than those found from the McMillan equation (Eq. 2-6b), which is constructed from numerical solutions of the Eliashberg equations for lower λ_{ep} values.

To solve the Eliashberg equations, values of μ^* (Coulomb-repulsion term, taken as $\mu^* = 0.1$) and $\alpha^2 F$ (Section 2-6), where F (ω) is the phonon density of states, must be known. Two approximations for $\alpha^2(\omega)$ were taken

$$\begin{aligned} \alpha^2 &= c_F & \text{(F — flat)} \\ \alpha^2 &= c_L \omega & \text{(L — linear)} \end{aligned} \qquad (5-6d)$$

In the first equation, α^2 is assumed to be flat or independent of frequency (F), and in the second it depends linearly (L) on frequency. The calculations were done for three representative crystals, 2-Bi(n=2), Y123, and La(n=1), which are labeled Bi, Y, and La, respectively. For these three

materials, $\alpha^2 F(\omega)$ is shown in Figs. 5-10a and 5-10b for the two approximations, and these can be compared to the generalized phonon density of
states in Figs. 5-9a and 5-9b.

Some results from these numerical solutions of the Eliashberg
equations are shown in Figs. 5-10c and 5-10d, where the gap-to-T_c ratio vs.
λ_{ep} and T_c vs. λ_{ep} are plotted. As can be seen, for $\lambda_{ep} \approx 2.6$, $T_c \approx 100$ K, and
$2\Delta(0)/k_B T_c \approx 5.4$. For conventional superconductors, relatively large λ_{ep}
values are found in the Pb-Bi system with tunneling data in amorphous-
$Pb_{0.45}Bi_{0.55}$ yielding $\lambda_{ep} = 2.59$. Thus, if high-T_c materials are strong-
coupled superconductors, $\lambda_{ep} = 2.6$ is not out of the question (also see
Section 5-6f), and these results of $T_c \approx 100$ K and a gap-to-T_c ratio of 5.4
are in reasonable agreement with experimental high-T_c results.

Another interesting feature of these results is that T_c keeps increasing
with increasing λ_{ep} (Fig. 5-10d). This increase is much stronger than those
from the McMillan equation (Section 2-6c), which tends to saturate in a T_c
vs. λ_{ep} plot (Notes).

5-7 Magnetic Properties

5-7a Type II Materials — An external magnetic field is excluded
from a type I superconductor for $H < H_c$ (the **critical field**). For H above
H_c, the field penetrates the entire sample and superconductivity is completely destroyed (Fig. 2-1). On the other hand, for a type II superconductor, the magnetic field is excluded only for $H < H_{c1}$ (**lower critical field**).
For $H > H_{c1}$, the field penetrates in vortices, and the fluxoid associated
with each vortex is one fluxoid quantum Φ_0 (Section 2-5d). As the external
magnetic field is increased, the density of vortices increases until the **upper
critical field** H_{c2} is reached, whereupon the field penetrates uniformly and
the material becomes normal (Fig. 2-5a). Values of H_c (Fig. 2-1) and H_{c1}
tend to be small ($<10^3$ Oe). However, H_{c2} values can be much larger (Fig.
2-5b).

All high-T_c materials are type II superconductors. The behavior of
vortices, pinning, and related properties are discussed in Chapter 6. In this
section, more macroscopic aspects of the type II, high-T_c superconductors
are considered.

5-7b Penetration Depth — Below H_c or H_{c1}, the external magnetic
field is excluded from the bulk of the material by a persistent supercurrent
in the surface region, which induces a field that exactly cancels the applied

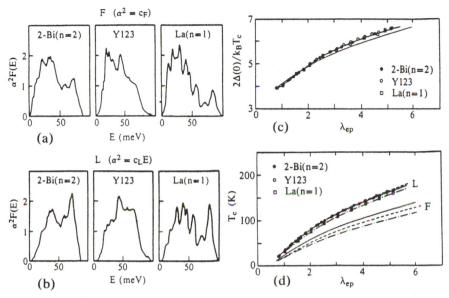

Fig. 5-10 (a) $\alpha^2 F$ vs. E, assuming α^2 is independent of energy, for the three superconductors as indicated. (b) is the same but assuming $\alpha^2 = c_L E$ (i.e., α^2 is proportional to energy). (c) The gap-to-T_c ratio vs. λ_{ep}. (d) T_c vs. λ_{ep} results using the flat and linear approximation for α^2. For details, see Y. Shiina and Y.O. Nakamura, Solid State Commun. **10**, 1189 (1990).

field. The depth of this surface layer is called the **penetration depth**, λ, and the external field penetrates the superconductor in an exponentially decreasing manner (Eq. 2-3e). The temperature dependence of λ was first calculated with the two-fluid model, which yields

$$\lambda(T) = \lambda(0)\,[1 - (T/T_c)^4]^{-1/2} \qquad (5-7a)$$

where $\lambda(0)$ is the value at absolute zero (Section 2-3). Figure 5-11a is a schematic diagram of the behavior of $\lambda(T)$, where it can be seen that λ (and also the coherence length) become very large near T_c. Moreover, for the highly anisotropic cuprate superconductors, λ is expected to be anisotropic.

$\lambda(T)$ can be directly measured for $H < H_c$ or H_{c1}. The usual growth habit of Y123 yields crystals that are much larger in the **ab**-plane directions than in the **c**-axis direction, which we take to have a thickness $=d$. For a magnetic field parallel to the **ab** plane (perpendicular to the **c** axis), the shielding current, which circulates about the field direction, mostly flows in the **ab**-plane directions in the crystal to a depth λ. The susceptibility $\chi\ (\equiv M/H)$ is then given by:

$$\frac{\chi(T)}{\chi_0} = 1 - \frac{2\lambda}{d} \tanh(d/2\lambda) \qquad (5-7b)$$

where χ_0 is the zero-temperature susceptibility and the hyperbolic tangent factor is a correction term close to unity for $2\lambda/d \ll 1$. This equation describes how the susceptibility is reduced by the volume fraction of the sample penetrated by the field; the factor 2 arising from penetration on both sides of the sample.

Figure 5-11b shows the result for $\lambda_{ab}(T)$, where the subscript ab refers shielding currents flowing in the **ab**-plane direction (as in the geometry discussed before). $\lambda_{ab}(T)$ rapidly increases as T_c is approached from below and the temperature dependence is in agreement with Eq. 5-7a, with $\lambda_{ab}(0) = 1400\text{\AA}$.

BCS (s-wave pairing) predicts a temperature-dependent penetration depth similar to Eq. 5-7a, with small changes in the strong-coupled limit, and small changes in the dirty limit. On the other hand, for p-wave and d-wave pairing (Section 5-2d), a rather different temperature dependence is calculated and shown by the straight line in Fig. 5-11b. Thus, the temperature-dependent penetration-depth results for Y123 (and other high-T_c superconductors) give strong weight to the argument that these materials are s-wave, BCS-type superconductors (Section 2-5c).

5-7c H_{c1} — Both H_{c1} and λ are closely related theoretically and in the way that they can be measured. Figure 5-11c shows results of Y123 crystals for H_{c1} parallel and perpendicular to the **c** axis. These results were obtained by placing Y123 single crystals in a very small coil that forms part of the tank circuit of an oscillator. As an appropriately oriented external field is slowly varied, oscillator-frequency changes are measured. Penetration of vortices at H_{c1} causes a sharp kink in the frequency vs. H curve. Since the results depend on a frequency measurement, high accuracy is obtained. The solid lines in Fig. 5-11c are BCS fits to the data using a gap ratio of $2\Delta/k_B T_c = 4.3$ for both the parallel and perpendicular directions. The dashed lines are for weak-coupled BCS (i.e., 3.5 for the gap-to-T_c ratio), which does not fit the data quite as well. However, the BCS-like character of H_{c1} for both directions is clear. Note that the H_{c1} anisotropy is approximately temperature-independent.

Y124 (Section 3-2g) is structurally related to Y123; the difference is in the details of the chains. For Y124, the temperature dependence of H_{c1} data has a shape similar to that shown in Fig. 5-11c. However, for **H** parallel to **c**, $H_{c1}(0) = 200$ Oe and for **H** parallel to either **a** or **b**, $H_{c1}(0) = 45$ Oe. Thus, H_{c1} shows no measurable anisotropy in the **ab** plane of Y124.

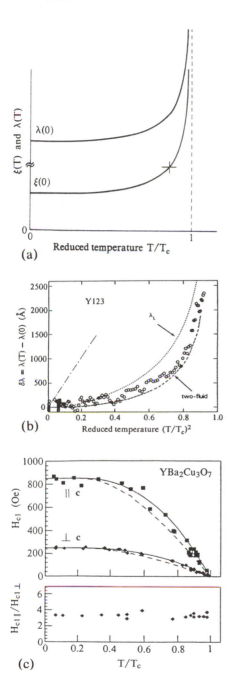

(a)

(b)

(c)

Fig. 5-11 (a) A schematic diagram of the temperature dependencies of the penetration depth (λ) and coherence length (ξ). The absolute values of these quantities can be quite different; in fact, for the high-T_c materials $\lambda \gg \xi$. However, according to the Ginzburg-Landau theory (Section 2-4), near T_c they have the same temperature dependence. (b) The open circles are measured changes in magnetic penetration depth λ as a function of reduced temperature. The dashed line is calculated using Eq. 5-7a with $\lambda(0) = 1400\text{Å}$ (two-fluid), and the dotted line represents weak-coupled BCS (λ_L), also with $\lambda(0) = 1400\text{Å}$. See L. Krusin-Elbaum, R. L. Greene, F. Holtzberg, A. P. Malozemoff, and Y. Yeshurun, Phys. Rev. Lett. **62**, 217 (1989). The straight, dash-dot, line is experimental data from the heavy-electron superconductor UBe_{13}; see D. R. Harshman, et al., Phys. Rev. B **36**, 2386 (1987). The latter result is from a non-s-like superconductor and the difference is apparent. (c) H_{c1} for H parallel and perpendicular to the **c** axis of single crystal Y123. The solid and dashed lines are discussed in the text. The 0 K values are 850 and 250 Oe. See D. H. Wu and S. Sridhar, Phys. Rev. Lett. **65**, 2074 (1990). Similar experimental results have been found by L. Krusin-Elbaum, et al., Phys. Rev. B **39**, 2936 (1989), who conclude that their results are in agreement with weak-coupling BCS (in the clean limit).

For an isotropic superconductor, the lower critical field (H_{c1}) is related to the penetration depth by

$$H_{c1} \approx \Phi_0/4\pi\lambda^2 \qquad\qquad (5-7c)$$

where Φ_0 is the fluxoid quantum (Eq. 2-5d). Thus, H_{c1} can be determined from λ or independently measured as discussed. The anisotropic character of λ and H_{c1} are discussed in Section 5-7e.

5-7d Coherence Length and H_{c2} — Besides having T_c values that are extremely high, these new materials also differ markedly from the conventional superconductors by having extremely small coherence lengths. The coherence length ξ is essentially the spatial range, or the decay distance, of the superconducting wavefunction. One may also think of it as the average diameter, or range, of a Cooper pair. For conventional superconductors, ξ typically varies from ~500Å to 10^4Å.

An important dimensionless parameter is the ratio of the magnetic penetration depth to the coherence length, the **Ginzburg–Landau parameter** $\kappa \equiv \lambda/\xi$. Abrikosov showed (1957) that, if $\kappa > 1/\sqrt{2}$, type II superconductors result (Section 2-5f). For most elemental superconductors, the GL parameter $\kappa \ll 1$, and, therefore, they are type I superconductors. However, essentially all of the compound conventional superconductors, and certainly all high-T_c materials, are type II superconductors. In fact, since κ of high-T_c materials is ~100, as we shall see, they are described as in the **extreme type II limit**, and they are in the clean limit, since ξ is much smaller that the electron mean free path (100Å to 200Å).

Direct measurements of ξ are difficult. The coherence length can be extracted from fluctuation contributions to the specific heat, susceptibility, or conductivity. However, for the high-T_c materials, reliable values of the coherence length have only been obtained via H_{c2}.

For type II superconductors, the coherence length is related to H_{c2} (similar to Eq. 5-7c),

$$H_{c2} \approx \Phi_0/2\pi\xi^2 \qquad\qquad (5-7d)$$

where Φ_0 is the fluxoid quantum. As can be seen, small coherence lengths mean large H_{c2} values. The problem is that these materials have huge H_{c2} values (greater than 100 tesla, where 1 T = 10^4 Oe) at absolute zero, so that H_{c2} can be reliably measured only near T_c. Some results are shown in Fig. 5-12 for a single-domain Y123 crystal along all three orthorhombic axes. The data were taken by measuring magnetization, M vs. T for different applied fields (H). The onset temperature of superconductivity for the

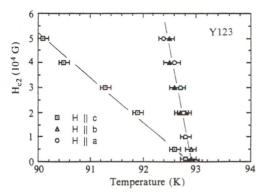

Fig. 5-12 Experimental results of H_{c2}^a, H_{c2}^b, and H_{c2}^c in single-domain Y123 crystals. Within experimental error the former two give the same result. For details, see U. Welp, M. Grimsditch, H. You, W. K. Kwok, M. M. Fang, G. W. Crabtree, and J. Z. Lin, Physica, C**161**, 1 (1989).

different H values is determined by a kink in the M vs. T curve, with the field corresponding to this onset temperature taken as H_{c2}. Note that H_{c2} along the **a**- and **b**-axes are essentially the same. From these measurements, $H_{c2}(0)$ can be found by using the **Werthamer Helf and Hohenberg** dirty-limit formula,

$$H_{c2}(0) = 0.693 \left| \partial H_{c2}/\partial T \right|_{T_c} T_c \qquad (5-7e)$$

H_{c2} at 0 K are estimated to be 670 T and 120 T in the **ab** plane and along the **c** axis, respectively, remarkably large values. Once $H_{c2}(0)$ parallel and perpendicular to the **c** axes are found, then the coherence lengths can be estimated using Eq. 5-7d. At 0 K, $\xi_c(0) \sim 3\text{Å}$ and $\xi_{ab}(0) \sim 16\text{Å}$ are found. These are remarkably small values. The anisotropy is discussed more carefully in Section 5-7e.

   Summarizing these and other quoted values at 0 K, the coherence lengths parallel to the **c** axis are typically 2-5Å, and in the **ab** plane, the results are typically 10-30Å. Thus, perpendicular to the **ab** plane, the superconducting wavefunction is essentially confined to the immediately adjacent Cu-O planes at low temperatures. Even parallel to the planes, it extends only a few unit cells. Such small lengths mean that the coherence volume contains only a few Cooper pairs, implying that fluctuations may be orders of magnitude larger in the high-T_c superconductors than in the conventional ones. Also, these very small ξ values imply a greater sensitivity of the superconductor properties to small-scale chemical and structural imperfections. Since these materials have complicated chemistry and structures (Chapter 3), this could be a serious problem. Undoubtedly, vortex pinning (Section 6-2b) is affected by the small coherence lengths. Remember, however, that ξ does become very large as T_c is approached (Fig. 5-11a).

5-7e Anisotropic Ginzburg-Landau Results — The GL theory (Section 2-5) has been extended to cover anisotropic materials. This is done with the relatively simple conceptual approach of replacing the mass by an **effective-mass tensor**.

For an orthorhombic or tetragonal crystal, the principal axes of a second-rank tensor lie along the crystallographic axes. The principal axes of the effective mass tensor have elements m_i, where i=1, 2, 3 corresponding to the **a**, **b**, and **c** axes, respectively, and the elements are normalized such that

$$m_1 m_2 m_3 = 1 \qquad (5-7f)$$

Penetration depths λ_i and coherence lengths ξ_i can be written in terms of the normalized masses as

$$\lambda_i = \lambda(m_i)^{1/2} \qquad \xi_i = \xi/(m_i)^{1/2}$$
where $\qquad \lambda = (\lambda_1\lambda_2\lambda_3)^{1/3} \qquad \xi = (\xi_1\xi_2\xi_3)^{1/3} \qquad (5-7g)$

Then, the GL parameter $\kappa=\lambda/\xi$, as usual (Notes).

The coherence lengths can be related to values of the upper critical field H_{c2i} measured along the three principal directions in a manner similar to Eq. 5-7d,

$$H_{c2a} = \Phi_0/2\pi\xi_b\xi_c \qquad H_{c2b} = \Phi_0/2\pi\xi_a\xi_c$$
$$H_{c2c} = \Phi_0/2\pi\xi_a\xi_b \qquad\qquad\qquad (5-7h)$$

Although these relations should be valid at all temperatures, remember that the GL theory is expected to be valid only close to T_c. Since ξ_c is the smallest coherence length, when it becomes smaller than c-axis repeat distance, one must be careful.

Expressions for an isotropic lower critical field H_{c1i} in terms of these parameters are more complicated because of mixing of λ_i and ξ_i terms. For tetragonal symmetry, we use the symbols parallel and perpendicular for the c-axis and the **ab**-plane directions. Then

$$H_{c1\parallel} = [\Phi_0/4\pi\lambda_\parallel^2] \, [\{ \ln(\lambda_\parallel^2/\xi_\parallel^2)\} + 0.5]$$
$$H_{c1\perp} = [\Phi_0/4\pi\lambda_\parallel\lambda_\perp] \, [\{ \ln(\lambda_\parallel\lambda_\perp/\xi_\parallel\xi_\perp)^{1/2}\} + 0.5] \qquad (5-7i)$$

As an example of the use of the anisotropic GL theory, we apply it to orthorhombic Y123. Using **Bitter pattern (magnetic) decoration experiments**, ratios of the penetration depths have been obtained,

$$\lambda_a : \lambda_b : \lambda_c = 1.15 : 1 : 5.5 \qquad (5-7j)$$

Note that these decoration experiments reveal an anisotropy between the **a** and **b** axes that is not found in the λ and H_{c1} experiments. From these penetration-depth ratios, the effective mass ratios can be obtained from Eq. 5-7g and are

$$m_a : m_b : m_c = 0.39 : 0.29 : 8.78 \qquad (5-7k)$$

Thus, the mass along the **c** axis is very much larger than that in the **ab** plane. These results also point to a coherence length that should be slightly anisotropic in the **ab** plane, as found for λ in Eq. 5-7j.

GL masses and band masses — The term **band mass** or **effective mass** refers to the usual m* that is extracted from band theory (Section 4-8) or obtained from E vs. **k** measurements. For some well studied semiconductors such as GaAs, InAs, and other III-V semiconductors both the top of the valence band and bottom of the conduction band have isotropic (independent of **k** direction) band masses. In both Si and Ge, the valence bands have isotropic band masses but the conduction bands have anisotropic mass such that the mass along one **k** direction is different from that along the two perpendicular directions. However, the high-T_c crystals have very complicated E vs. **k** behavior (Section 4-8b); near $\overline{M}$, $E(k)$ is not quadratic, so an effective mass can not be extracted. Thus, the question arises, what is the relation between GL superconducting masses (Eq. 5-7f) and band masses? Since the GL masses have no units and are normalized to unity, the question should be rephrased in terms of the meaning of the ratios. The answer is that there is very little relationship, except at the deepest, and most difficult to calculate, theoretical level. For example, H_{c2} has been calculated in terms of microscopic properties of some conventional superconductors. Rather, the GL masses are phenomenological mass ratios that can be used to relate the GL length scales, and they can be related to some other macroscopic measurements (Section 5-7f).

T$'$ structure — H_{c2} measurements of $Nd_{2-x}Ce_xCuO_4$ with $T_c \approx 22$ K have been made. This material is the electron-doped high-T_c superconductor with the T$'$ structure (Section 3-2g). Extrapolated 0 K values are 6.7 T and 137 T along the **c** axis and in the **ab** plane, respectively. These results correspond to coherence lengths (Eq. 5-7h), $\xi_c \approx 3.4$Å and $\xi_{ab} \approx 70$Å. Thus, the results are similar to those in Y123.

Perovskite superconductors — For the $(Ba,K)BiO_3$ superconductors (Section 3-2g), an isotropic coherence length is found, $\xi(0) \approx 37$Å. Thus, the coherence volume is almost two orders of magnitude larger than in Y123, again emphasizing the difference between these and the planar-cuprate superconductors.

Dimensional-crossover effects are interesting to consider in Y123 because the materials are very anisotropic, yet superconductivity is essentially a three-dimensional phenomenon. At 0 K in Y123, we have $\xi_c(0) \sim 3\text{Å}$. The two immediately adjacent Cu-O planes are $\sim 3.2\text{Å}$ apart, but each of these two immediately adjacent planes is more than 8Å away from one of the next two planes in Y123 (Chapter 3). Thus, at ≈ 0 K, the coherence length along the **c** axis is equivalent to the separation between the immediately adjacent Cu-O planes but considerably smaller than 8Å. Therefore, in some sense, the superconductivity might be thought of as two-dimensional. On the other hand, for a phase transition, ξ must become very large near T_c, as indicated in Fig. 5-11a. A **crossover temperature**, T^+, can be defined, which might be considered to correspond to a transition temperature from high-temperature, three-dimensional behavior to lower-temperature, two-dimensional behavior. Using the Ginzburg-Landau theory for a specific temperature dependence of $\xi(T)$, then $\xi_c \sim 8\text{Å}$ at ~ 70 K. Thus, $T^+ \approx 70$ K, which is denoted by the cross in Fig. 5-11a.

5-7f   Torque Magnetometry — If a magnetic field is applied to a superconducting cuprate where **H** is out of the conduction plane, then there is a torque on the sample. The thermodynamic potential for different crystal orientations with respect to **H** can be calculated in terms of the anisotropic GL masses (Notes).

Accurate torque-magnetometry measurements have been made for many of the high-T_c materials. For 2-Tl(n=2), an anisotropy of $\sim 10^5$ is found for the ratio of the mass along the **c** axis to that in the **ab** plane. A similar large ratio is found in 2-Bi(n=2). These extremely large ratios are much larger than the ratio of ~ 25 found in Y123 (Eq. 5-7k). These large mass ratios emphasize some aspects of the two-dimensional character of the high-T_c superconductors.

5-8 Postscript

In spite of the extensive, apparently good experimental results on high-T_c materials, it is surprising how few noncontroversial facts have resulted. When some experimental point becomes apparently settled, within a relatively short time other experimental results lead to different interpretations. In many respects, opportunities for clean, definitive experiments appear to be as strong now as in the early high-T_c years. Also, after decades of research, only the BCS model is well developed. It would be nice if some of the other, and particularly the more recent, models of nonconventional superconductivity could be developed far enough to allow detailed comparison to more experiments.

Problems

1. Debye solid — For the Debye model, show that $\Theta_D \propto M^{-\frac{1}{2}}$.

2. Phonons plus electron density of states singularity — Using Eq. 5-6c, calculate T_c when E_F is just at the van Hove singularity. How does T_c vary with the Fermi temperature; for T_c to decrease by two, what change must occur in T_F?

3. Spin selection rules — The spin part of the spin-singlet wavefunction in Eq. 5-2d can be written as

$$(\alpha\beta - \beta a)/\sqrt{2}$$

where the electron numbers are implicit in the order of writing the spin functions. Electric dipole transitions are governed by matrix elements of the electric dipole operator, $\mu = er$. (a) Show that transitions between a singlet and triplet are **spin forbidden** by considering the transition between the spin singlet and the $\alpha\alpha$ component of the spin triplet. (b) Show that a singlet-singlet transition is spin-allowed. (c) For a triplet-triplet transition, which components of the triplet make the transition spin allowed? (d) Why are the above conclusions also appropriate for electric quadruple transitions?

Chapter 6

Vortex Behavior, J_c, and Applications

*Fire comes and the news is good, it races through the streets
but is it true? Who knows? Or just another lie from heaven?*

Aeschylus, "Agamemnon"

6-1 Introduction

The words "superconducting state" are usually associated with a state
of matter in which the dc electrical resistivity is zero, or as close to zero as
clever experimentalists can measure. This is true in sufficiently weak magnetic fields. In fact, decay of **persistent currents** in superconducting rings
have been measured; from this data and calculations, the decay time has
been estimated to be longer than the age of the universe. However, all of
this is only true for type I superconductors or type II superconductors in
sufficiently weak magnetic fields, which, thus, are vortex-free.

As discussed in Section 2-5e, the Ginzburg-Landau (GL) parameter
κ is the magnetic penetration depth divided by the coherence length
($\kappa \equiv \lambda/\xi$). For most elemental superconductors, $\kappa < 1$ because $\lambda < \xi$.
Abrikosov showed that for $\kappa > 1/\sqrt{2}$, what is called a **mixed state**, or
vortex state, is formed. In this state, for magnetic fields, $H_{c1} < H < H_{c2}$,
tubes of flux-bearing (normal-state) material occur in the superconductor.
A tube of material is called a **vortex**, or **filament**, and at its center the
superconducting order parameter reaches zero. Along principal axes, the
tube is parallel to **H**, and current circulating around a vortex produces a
magnetic field that screens the normal-state vortex core from the superconducting region outside. *These circulating currents are not transport currents*; rather, they are similar to the surface currents that shield the bulk of
a type I superconductor, yielding perfect diamagnetism.

(a)

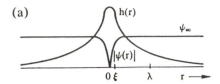

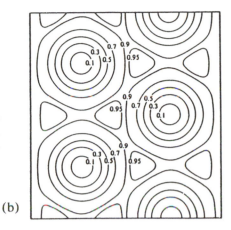

Fig. 6-1 (a) Parameters near an isolated vortex. (b) A calculated spatial distribution of $|\psi|^2$ for a vortex array near H_{c2}. Note the hexagonal lattice. See W. H. Kleiner, L. M. Roth, and S. H. Autler, Phys. Rev. B **133**, 1227 (1964).

(b)

Figure 6-1a shows the radial dependence of the superconducting order parameter $\psi(r)$ and the local microscopic field $\mathbf{h}(r)$ for an isolated vortex. (See Section 2-5b for a discussion of the local microscopic field.) This figure is drawn with $\kappa \sim 8$. As can be seen, the order parameter reaches its $r=\infty$ value (ψ_∞), for $r > \xi$. The maximum value of $\mathbf{h}(r)$ is approximately $2H_{c1}$. Circulating around the vortex center are shielding currents such that each vortex carries a total flux equal to the fluxoid quantum $\Phi_0 = hc/2e$ (Section 2-5d).

For $H_{c1} < H < H_{c2}$, barring various types of obstructions (pinning) and anisotropies, there is a distribution (an array) of cylindrical vortices whose axes are parallel to the applied field. This distribution forms a *regular array* in the form of a hexagonal lattice (a lattice is a periodic array), and a calculated result is shown in Fig. 6-1b for H near H_{c2}. The closeness of the vortices in Fig. 6-1b occurs because H is near H_{c2}, the opposite limit of the situation in Fig. 6-1a.

All high-T_c superconductors are type II. Thus, for $H_{c1} < H < H_{c2}$, they are in the vortex state, and this range of fields typically is required for most applications. Therefore, the issues raised above are critical for the understanding and use of these materials.

The problem — A calculation of, for example, the magnetoresistance (R vs. H) of a normal metal is straightforward because the field is uniform in the sample. Similarly, in a type I superconductor, as long as the penetration depth is small compared to sample dimensions, the resistance ($=0$) is understood because of the uniform field in the bulk ($=0$). However, in a type II superconductor, this uniformity is absent on a macroscopic and microscopic scale, and resistance can be measured in many situations. Thus, for type II superconductors, the situation is complicated and some general questions arise.

1. What is the nature of the vortex lines? Along a principal axis, are they parallel to each other, forming a hexagonal lattice? Are they parallel to each other but at random positions? Are vortices straight or do they have wiggles, perhaps due to being pinned at different sites along their lengths? Or are the vortices closer to being totally entangled, like cooked spaghetti on a plate?

2. What is meant by the critical current of a type II superconductor? How does it depend on the nature of the vortex state and how is it related to temperature and field?

3. How is E vs. J affected by the vortices and their motion; how does it depend on T and H? What contributes to the dc and ac resistance in the sample and how is it affected by the surface impedance?

The behavior of type II superconductors depends in a complicated manner on the issues brought up in these three points. Most models of their behavior are highly simplified, addressing different parts of these general problems.

In this chapter, the first few sections are devoted to elementary properties and vocabulary building. Then, related physics is addressed in Section 6-4. The behavior of vortices in large transport currents and high magnetic fields is a topic of current research, and this chapter is only a brief introduction to this area.

6-2 Flux Lattice, Flux Glass, and Pinning

6-2a Flux Lattice and Glass — The current circulating around a vortex (or filament) makes each vortex act like a tiny bar magnet. The vortices in a type II superconductor repel each other. Repulsion occurs because the currents circulating about one vortex produce a repulsive Lorentz force, $J \times B$, on the magnetic field of another vortex. If other forces are weaker than the vortex-vortex repulsion, then the vortices form a hexagonal, close-packed lattice, called a **flux lattice**, or an **Abrikosov lattice**. This lattice is observed in conventional as well as high-T_c superconductors. However, even in some conventional superconductors, irregular arrays of vortices are observed.

For Y123 single crystals, Abrikosov lattices are observed at low temperatures on carefully prepared single crystals cooled in a small field.

However, generally such patterns are not observed in good Y123 films; instead, a random flux pattern is most often found. Such a pattern is sometimes called amorphous or a **flux glass**, where the word "glass" refers to the randomness of the spatial pattern. In fact, for other high-T_c single crystals as well as films, flux glasses are almost always found rather than flux lattices. For $H_{c1} < H < H_{c2}$, the vortices may be **pinned** in their positions by forces that are large compared to the vortex-vortex force. Since the pinning centers are generally randomly distributed, a flux pattern determined by the spatial arrangement of the pinning centers occurs. Apparently, this is the case in Y123 films, leading to a flux glass (Section 6-4).

The problem — For a type II superconductor with $H_{c1} < H < H_{c2}$, it might be thought that the resistance is zero because current can flow through the superconducting regions avoiding the normal-state material at the vortex core. However, the situation is more complex. When a transport current passes through a type II superconductor for $H_{c1} < H < H_{c2}$, the Lorentz force ($\mathbf{J} \times \mathbf{B}$) acts on the vortices, since each vortex acts as a tiny bar magnet. If this force moves the vortices, then a longitudinal potential gradient parallel to the transport current develops, which is equivalent to a power loss, or a resistance in the material. Thus, *when vortices move under the influence of a transport current, energy is dissipated and the material has a resistive loss.* Resistanceless operation of type II superconductors for $H_{c1} < H < H_{c2}$ can be achieved only if the vortices are strongly pinned. Thus, methods to pin the vortices are important if the superconductor is to be useful for high-current applications. We return to the critical issue of resistance in Section 6-4.

6-2b Pinning — To pin vortices, potential wells must be created in which the vortex has a lower energy than in its surroundings. Thus, the vortex "sits" at a pinning site. If the potential well is very deep, neither the Lorentz force nor thermal energy ($k_B T$) will be able to move the vortex. Thermal energies are a particular problem in high-T_c materials because operating conditions may be at temperatures much larger than for conventional superconductors. Pinning results from spatial inhomogeneities of the material, such as impurities, grain boundaries, voids, dislocations, non-superconducting precipitates, and so forth. Pinning at these types of sites occurs because it costs less energy if the vortex core is at a position that is normal than if it must change a superconducting region to a normal region. To be effective, the inhomogeneities may be on the scale of λ to produce magnetic pinning. For high-T_c superconductors, the more important situation is vortex-core pinning, in which case the inhomogeneities should be on the scale of ξ. For high-T_c materials, ξ values are very different from those found in conventional superconductors, which leads to uncertainties in the approach and understanding of ways to pin vortices.

The very small ξ values in the high-T_c superconductors may also allow pinning to be accomplished by atomic defects; this may be especially true if the defects are on the Cu-O planes.

For conventional superconductors, the metallurgical procedures used to produce deep pinning potentials are similar to those used to obtain high-yield-strength materials (which tends to cause the loss of ductility). Defects of a certain type, size, and density are introduced into the structure. These defects include impurities, precipitates, defects formed by neutron or proton bombardment, dislocations (introduced by cold working), and grain boundaries. The science and art of producing pinning sites in commercially used Nb-Ti alloy wire is well studied. Commercially used Nb_3Sn and V_3Ga intermetallics are inherently brittle and the **bronze process** has been developed to make usable wires. In this process, Nb rods are inserted into Sn tubes, which are repeatedly drawn down. The resultant rod is inserted into Cu tubes and again repeatedly drawn down. The resultant wire is then heated (sometimes in its final configuration) so that the A15 compound is formed at the Nb-Sn interface. These wires have the advantage of being surrounded by Cu sheaths, which provide mechanical and current stability for the superconductor wires. For high-T_c materials, studies to make commercial wire or tape are in the beginning stage, and the methods may be quite different from those found in conventional superconductors because the metallurgy of these classes of materials is very different.

Y123 material grown so as to yield Y_2BaCuO_5 inclusions appears to have a pinning energy higher than pure Y123. This and related approaches are being studied. However, it is felt that single oxygen vacancies might also be pinning sites.

6-3 Films and Critical Currents

6-3a Films — Excellent Y123 films can be grown by several deposition techniques onto a substrate of, for example, single-crystal $SrTiO_3$ (or $LaGaO_3$, etc.). For the deposition, typically a Y123 pellet is evaporated by a high-power laser (often a pulsed excimer laser), or it is sputtered. The substrate, typically a (100) face of $SrTiO_3$, is maintained at $\sim 700°C$. An oxygen atmosphere during growth is preferable and the film properties may be improved by ionizing some of the O_2, which increases the chemical reactivity. In situ anneals are commonly used to obtain films with higher critical currents. Film growth by CVD, MOCVD, MBE, and other techniques is also being studied.

The good Y123 films (typically, $\lesssim 1\mu m$ thick) have their **c** axis perpendicular to the (100) substrate face (i.e., **c** axis is the growth direction). Thus, the current in the plane of the film flows parallel to the Cu-O planes (the **ab** planes). However, Y123 films with the **c** axis in the plane of the film also can be grown. There is considerable effort to grow other high-T_c materials and use other substrates, such as silicon and other semiconductors, in order to be able to combine the two technologies.

6-3b Superlattices — With reasonable control of the film-growth techniques, films for other types of experiments are being grown. Borrowing ideas from the semiconductor field, Y123 superlattices have been made. Both Y123 and Dy123 have $T_c \approx 93$ K. Superlattice **c**-axis films have been grown with ~12Å of Y123 on ~12Å of Dy123, repeated to yield a superlattice with no degradation of the superconducting properties due to interfaces. Of course, 12Å is the unit-cell size in the **c** direction. Thus, the superlattice-film technique is viable down to unit-cell sizes (Notes).

Pr123 is an insulator and also isomorphic to Y123. Superlattices with a variable Y123 thickness and a 100Å layer of Pr123 have been grown and studied. The 100Å insulator layers are experimentally shown to be thick enough to insulate the superconducting Y123 layers from each other, with no degradation of the crystallinity. For such Y123/Pr123 superlattices, T_c vs. d_{Y123} is shown in Fig. 6-2a, where d_{Y123} is the thickness of the Y123 layers. The results imply that a Y123 layer with the height of one unit cell along the **c** axis (12Å, Section 3-2f) is superconducting. However, coupling between the unit cells is needed to achieve the bulk- T_c values.

These results support the point of view that the high-T_c materials are three-dimensional superconductors. This means that coupling along the **c** axis between the Cu-O planes is required to achieve the high values of T_c. Of course, the materials still have highly anisotropic properties. Also, it is of significance that a single (12Å) layer of Y123 appears to be a superconductor (Fig. 6-2a).

6-3c Wires — In the early high-T_c days, wires made of these materials seemed an unlikely prospect because of the brittleness. However, remarkable progress is being made. Wires of 2-Bi(n=3), with some Pb replacing the Bi, are being made in kilometer lengths with good critical cur-

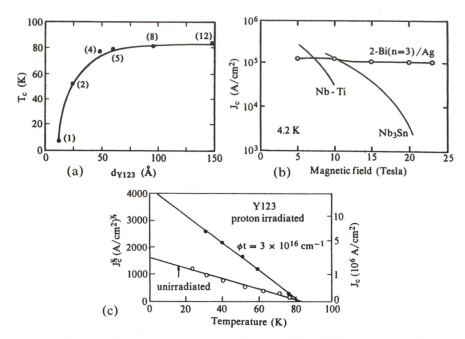

Fig. 6-2 (a) T_c vs. d_{Y123}, where d_{Y123} is the thickness of the Y123 layer in a Y123/Pr123 superlattice. (The Pr123 layer is 100Å thick.) The unit cell height along the c axis is ≈ 12Å, so the number of Y123 unit cells is indicated in the figure. For details, see J.M. Triscone et al., Phys. Rev. Lett. **64**, 804 (1990) and Q. Li et al., ibid, 3086 (1990). (b) J_c vs. H for two conventional (commercial) superconductors and silver-sheathed 2-Bi(n=3) wires. See the text for details and K. Sato, T. Hikata, H. Mukai, M. Ueyama, N. Shibata, T. Kato, T. Masuda, M. Nagata, K. Iwata, and T. Mitsui, Applied Superconductivity Conf. 1990 and IEEE Trans. Mag. **27**, 1231 (1991). (c) J_c vs. T of proton-irradiated Y123 single crystals in an applied field of 10^4 Oe (1 tesla). The lines are a guide to the eye and they indicate that empirically $J_c^{1/2}$ depends linearly on temperature in some crystals. For details, see R. B. van Dover, E. M. Gyorgy, L. F. Schneemeyer, R. J. Felder, and J. V. Waszczak, App. Phys. Lett. **56**, 2681 (1990).

rents. The wire production starts by filling a silver tube with the high-T_c material. The tubes are then drawn, rolled, and sintered, resulting in typical cross sections of ~0.1×2 mm. 2-Bi(n) is very micaceous, and the rolling and sintering causes grain alignment of the microstructure so that current in the wire flows mostly through **ab** planes with apparently minimal effects from weak links between grains.

Grain alignment in the 2-Bi(n) materials occurs much more readily than in Y123. Thus, at present, the former appears more promising. However, there are extensive efforts underway to form similar wires using Y123.

Methods of making wires and tapes of the high-T_c materials are in the early stages, but the rate of progress has been remarkably high. Preliminary critical current vs. magnetic field data for these 2-Bi(n=3) silver-clad wires is shown in Fig. 6-2b and compared to conventional superconductors at 4 K. For the 2-Bi(n=3)/Ag wires, J_c is not affected by the external field, while J_c at high fields is severely reduced for the conventional superconductors. The qualitative difference comes from the fact that H_{c2} (Fig. 2-5) is very much larger for 2-Bi(n=3) than for the conventional superconductors. It appears that very high field magnets made from 2-Bi(n=3) should be commercially feasible for operation at 4 K to 20 K. The latter temperature is significant, since only cryocoolers, rather than submersion in liquid helium, is required. Operation at 77 K using 2-Bi(n=3) will prove more difficult because of the weak pinning of the vortices found in this material; at 77 K, the pinning is much weaker than that found in Y123 (Section 6-4f).

6-3d   Critical Current — In his Nobel Prize lecture (1913), Kamerlingh Onnes discussed the possibility of superconducting magnets that would produce large magnetic fields yet consume no electrical power. Unfortunately, J_c values of type I superconductors are too small. It was not until the exploitation of type II superconductors (since the 1950s) that this has become practical.

The critical current, J_c, is the maximum current density that can be sustained by a superconductor before the particular measurement technique detects dissipation. However, care must be exercised because J_c is defined in different ways in different experiments.

Theoretically, the highest critical current density that a superconductor can carry is the **velocity depairing current** or the **depairing current density**. (The electron kinetic energy is equal to the gap energy.) Above this current, the superconductor is driven normal. The depairing current density can be estimated from the GL theory in type II superconductors near T_c to be (cgs units)

$$J_c = \frac{4}{3} e\psi_\infty^2 \left(\frac{2|\alpha|}{3m^*} \right)^{1/2} = \frac{cH_c(T)}{(3\sqrt{6})\pi\lambda(T)} \propto (1-t)^{3/2} \quad (6-3a)$$

where H_c is the thermodynamic critical field and $t=T/T_c$. (See the Notes and Problems.) For Y123, this calculated value is in the range 10^{8-9} at 4 K and 10^{7-8} A/cm^2 at 77 K. However, measured J_c values are often considerably smaller because of vortex motion.

Experimental determination of J_c is complicated because the resistive losses through a type II superconductor (for $H_{c1} < H < H_{c2}$) depend on the sample geometry and vortex pinning in the material. To define a critical current, a voltage criterion is usually used, which is simply related to Ohm's law,

$$J = E/\rho \qquad\qquad (6 - 3b)$$

For very small currents in a type II superconductor (in a magnetic field), the voltage required to sustain the current is proportional to the current ($V \propto I$), and the corresponding resistivity (Eq. 6-3b) is extremely low, at least within the framework of the Anderson-Kim theory (but see Section 6-4d). Then, as the current is increased, some value is approached where the voltage required to produce the current increases much more rapidly than given by Ohm's law. In this region, the I-V dependence is fitted to a power law, $V \propto I^n$. Usually, the critical current density is given by the current when this strong nonlinearity between V and I is found. For $n > 10$, the rise of voltage is so rapid that a fairly unique J_c can be determined, and this is usually the case.

If a current density of 100 A/cm^2 though a superconductor is produced by a 1 mV/cm electric-field drop, then $\rho = 10^{-8}$ Ω-cm, about that of Cu at 4 K. The ASTM (American Society for Testing Materials) has suggested that J_c be defined as the current density that results in a resistivity of 10^{-12}ohm cm. However, the I-V curves typically rise so rapidly that either definition yields similar J_c values.

J_c can also be estimated from magnetization measurements. The hysteresis in magnetism curves allows J_c to be determined with the help of the **critical state model**. This model assumes that, for $H > H_{c1}$, the critical current is flowing in the sample to prevent the penetration of flux. J_c determined from magnetic measurements and direct resistance measurements is usually in good agreement. The former does not require contacts, which is an advantage.

The high-T_c materials are all type II superconductors. At present, the best Y123 films have $J_c \sim 5 \times 10^7$ amp/cm^2 at 4 K for fields 0 to 1 T, the value at 77 K being about 10 to 50 times lower. A curious finding is that generally J_c values of good Y123 single crystals are about 30 times smaller than those in films at 4 K for the same fields. However, at 77 K, they are 100 to 1000 times smaller than the values found in good films. The reason for this result is not clear, but it is associated with details of flux pinning, which apparently is better in films than in single crystals.

Results of J_c vs. H (at 4.2 K) for Ag-sheathed 2-Bi(n=3) wires, discussed in Section 6-3b, are shown in Fig. 6-2b. As can be seen, J_c is essentially independent of field up to 25 T, while it decreases rapidly for Nb-Ti and Nb$_3$Sn. The reason for this behavior is that H_{c2} at 4 K is very much larger than 25 T for 2-Bi(n=3), while for the conventional superconductors, the $H_{c2}(T)$ line is being approached (Fig. 2-5), so J_c is reduced.

For many conventional superconductors, pinning sites can be created (increasing J_c) by **bombardment of the sample** with fast neutrons, protons, or other particles. The fast (high-energy) particles should have enough energy to penetrate the entire sample. This is important because at the end of most particle tracks, when the particle energy is largely dissipated, the cross section for damage increases dramatically, which would cause local amorphization of the sample.

J_c of Y123 single crystals can be improved by fast particle bombardment. Figure 6-2c shows results in Y123 of J_c vs. T for both an unbombarded and a proton-bombarded single crystal. The **fluence**, ϕt, for these ~3 MeV protons is determined by the number of particles per area impinging on the target (the flux) times the time of exposure. For the results shown in Fig. 6-2c, the flux could be 6×10^{12} cm^{-2} sec^{-1} (~ 10^{-6} A/cm^2) for 5000 sec; lower flux requires longer times for the same fluence. The results in Fig. 6-2c give the maximum improvement for that particular sample, the improvement peaking for the fluence given. Apparently, the proton bombardment creates small damage regions that pin vortices (probably at several places along its length) in the Y123 single crystal. At low fluence, J_c increases as the bombardment increases. However, at very high fluence, there may be extensive crystal damage, so J_c decreases or the damage centers begin to overlap, reducing the effective pinning barrier, or both. Thus, J_c vs. fluence should peak, as is experimentally found. Recent particle bombardment results have pushed J_c of Y123 crystals to about the same value as that found in the films.

Particle bombardment of good Y123 films improves J_c values only slightly. Apparently, the vortices are pinned as well, and at as many places as possible, by defects in the films. Thus, the J_c values in "good" Y123 films may be close to the upper limit that can be expected.

J_c of Y123 polycrystalline films, those with the **c** axis in random directions, are much smaller than those of the films discussed above. This occurs because in some grains of a polycrystalline film, the current must flow parallel to the **c** axis, and J_c parallel to the **c** axis is considerably smaller than J_c in the **ab** plane. Even in an **ab**-oriented film, the boundaries between different grains may act like weak links (Section 6-7), and thus, the measured J_c is reduced from the value found in a single grain.

The J_c values in the good films should make high-current-density devices practical. For example, these values can be compared with those found presently in commercial superconducting magnets made from Nb_3Sn and Nb-Ti, which, when operating at 4 K, have J_c values comparable to the good Y123 films when the latter are operating at 77 K. However, such devices typically require long lengths of superconductor material. Thus, the technology for depositing long lengths of high-current-density superconductor materials must be developed. Probably, such material will be deposited in the final device shape, since the high-T_c materials are not ductile.

For films, only those made from Y123 have been discussed because they have been the most extensively studied. However, it may turn out that other high-T_c materials will produce better films or wires (Section 6-3c).

6-4 Macroscopic Magnetic Properties

6-4a Introduction — From a broad point of view, there are two types of applications for superconductors. The first is **large-scale applications**, in which large currents and long lengths of superconductors are typically required in environments where the magnetic field may be several tesla (1 T $= 10^4$ Oe). Examples include magnets and power transmission lines, transformers, and generators where current densities of at least 10^5 amps/cm^2 are required. For many large-scale applications, the only advantage of superconductors over normal metals is lower resistance and hence smaller power loss. The second type of application is **small-scale applications**, where more specialized properties of superconductors tend to be exploited. Examples include SQUIDs, other detection systems, and analog and digital processing. Some small-scale applications are discussed later in this chapter.

To understand the physics behind the use of these superconductors for large-scale applications, the H vs. T phase diagram must be understood. For type I superconductors, the simple H-T phase diagram is shown in Fig. 2-1. For H=0, there is a second-order phase transition at T_c and if H>0, a first-order phase transition occurs at $H_c(T)$ given by the curve for a particular material. For type II superconductors, the H-T phase diagram is more complicated (Fig. 2-5a) because of the existence of the vortex state for $H_{c1} < H < H_{c2}$. This result (Fig. 2-5a) is a mean-field result that ignores both pinning and thermal fluctuations. *However, in the presence of pinning and thermal fluctuations, the phase diagram is more complicated and also controversial.* It is the type II results that are discussed here.

6-4b Vortex Glass — As mentioned in Section 6-2a, a (hexagonal) vortex lattice can be observed in both conventional and high-T_c superconductors under certain conditions. However, for most high-T_c films and crystals, and many conventional superconductors, the observed vortices have little or no long-range order; they are randomly arranged in a spatially amorphous manner. The randomness of this phase arises from the randomness of the pinning. In high-T_c materials, vortex pinning sites possibly include oxygen vacancy-type positions and other atomic disorder that may arise from slight deviations from the complicated stoichiometry, twin boundaries, and other deviations from translational symmetry. Since the coherence length is so small (3 to 30 Å), the vortex core is small (Fig. 6-1a); thus, pinning might be accomplished by inhomogeneities on the single-atom scale. For **H** in the **ab** plane, additional pinning appears to occur via the isolation planes separating the immediately adjacent Cu-O planes (Figs. 3-1 to 3-5). This is called **intrinsic pinning**.

6-4c Flux Creep — For any superconductor, pre-1986 thinking was that vortex motion can be divided into two types, the first usually being acceptable for applications and the second unacceptable.

1. Flux creep — If the thermal energy and Lorentz forces that arise from a transport current are smaller than most of the pinning energies (taken as U on the average), the vortex motion is very small and the resistance of the superconductor is extremely small and of little consequence. There will be a measurable amount of vortex hopping that may be thermally driven. Flux-density gradients are gradually reduced by vortex creep, which occurs at an orderly (and tolerable) pace under the influence of a transport current. Vortices jump from one pinning site to another, or a "bundle" of vortices jump. Jumping may happen in bundles, because the vortex-vortex forces are always operative. Flux creep is found immediately after the change of an external field or current. It can be detected by measuring the trapped magnetic fields or the resistance. This flux creep motion is found to decay logarithmically with time.

2. Flux flow — For a given pinning potential, as the transport current increases, the Lorentz force can become of the same order of magnitude as the pinning forces. Then the vortices move with a steady motion, their velocity being limited by vortex viscous-drag forces. This motion is the **flux**

flow state; the resistance is $\sim \rho_n H/H_{c2}$ where ρ_n is the resistance of the material in the normal state. This numerical value of resistance comes from the **Bardeen–Stephen model**. Once there is a flux flow and a resistance in the material, thermal heating may produce a positive-feedback mechanism to further enhance the flux flow.

In the flux-flow state, the material is of little use as a low-loss conductor. Hence, for practical applications, flux-flow must be avoided and flux creep minimized. Now we begin to explore a subtle question: does the material go from a flux creep to a flux-flow state in a gradual manner or is there a phase transition, at least in the limit of zero current?

6-4d A True Zero Resistance State? — Superconductivity implies a zero resistance state below T_c; for very small currents and $H < H_{c1}$, the resistance is zero. In fact, persistent current experiments set extremely small upper limits on the resistivity. However, for a type II superconductor with $H_{c1} < H < H_{c2}$, the situation is less clear because of the presence of vortices. There are two general points of view as to how the material changes from a flux-creep to a flux-flow state. The views are discussed here and experimental results are presented in Section 6-4e.

The linear resistance is defined in the limit of zero current as

$$R_L \equiv \lim_{I \to 0}(V/I) \qquad (6-4a)$$

The Anderson-Kim (1962) **flux–creep model**, and extensions thereof, include the effects of pinning of vortex lines. The flux-creep model predicts a resistivity,

$$\rho = (2\nu H L_p/J)\, e^{-U/k_B T}\, \sinh(JHV_c L_p/k_B T) \qquad (6-4b)$$

where ν is a flux-line attempt frequency (as in atomic diffusion), L_p is the jump distance of the flux line, and V_c is the vortex-bundle volume. The barrier energy for a vortex jump (or a bundle of vortices) is U. At low current, the hyperbolic sine term can be replaced by its argument, so J cancels out of the expression. That results in a *current independent resistivity*, which has an experimental temperature dependence, $\exp - (U/k_B T)$. Thus, this model predicts a finite resistivity at all temperatures (except at zero), and ρ has a thermally activated form.

More recently, the **vortex–glass model**, or Fisher model (1989), takes into account both pinning and the collective effects of the vortex lines. In this model, a single vortex line may be composed of many segments pinned with different energies, and the collective behavior of these segments as well as the collective behavior of the many vortex lines is taken into account.

As mentioned, the repulsive vortex-vortex forces oppose independent vortex motion. An important prediction of this model is that, below a certain temperature $T_g(H)$, the vortex system will freeze into a **vortex–glass phase** in which, indeed, $R_L = 0$. Above $T_g(H)$, the vortex system is a **vortex liquid** (the flux-flow state) with a resistance not very different from that of the normal-state metal. Further, this theory predicts a second-order phase transition between the vortex-glass phase and the vortex-liquid phase for all values of H. Thus, the H-T phase diagram is conceptionally quite different from mean-field results (Fig. 2-5a) and from the flux-creep model.

6–4e Experimental Vortex Glass–Liquid Measurements — To determine the resistance of the vortex state in the limit of $I \to 0$, measurements at very low current densities must be performed. Figure 6-3a shows E vs. J results in a field of 4 tesla for a Y123 film (on $SrTiO_3$) with the c axes predominantly perpendicular to the substrate; hence, J is in the **ab** plane. Such films have $J_c \gtrsim 10^6$ A/cm^2 at 77 K. The high-temperature isotherms are in the upper left and the low-temperature ones in the lower right as indicated by T=80.8 and 67.3 K; the isotherms are shown in 0.3 K temperature intervals. (0.1 K was used for more accurate data.) The dashed line at T$\approx$77.3 separates the qualitatively different behaviors. At higher temperatures, as J is reduced, E is reduced less, implying R_L remains finite as $J \to 0$. However, for $T \lesssim 77.3$, the curves bend the other way, implying $R_L \to 0$ as $J \to 0$. Thus, $T_g = 77.3$ K at 4 T for this Y123 film.

Similar measurements at other magnetic fields yield $T_g(H)$, shown in Fig. 6-3b. These results imply $R_L = 0$ below $T_g(H)$, the vortex-glass state, while above $T_g(H)$, the film is in the vortex-liquid state. Further analysis of the experimental results (Fig. 6-3a) indicates that there appears to be a second-order phase transition between the vortex-liquid and vortex-glass phases for the measured magnetic fields. This result is determined by noting that all the measured E vs. J data above T_g can be scaled onto a single universal E vs. J curve for different magnetic field values. Similarly, a single universal E vs. J curve can represent all of the data below T_g. The scaling factors (critical exponents) required to perform this data reduction are in good agreement with the theoretical predictions. These experiments strongly support the vortex-glass to vortex-liquid phase transition model. Furthermore, similar results found in Y123 single crystals and polycrystals (Notes) also support this vortex-glass, vortex-liquid melting model.

In Fig. 6-3b, the mean-field H_{c2} results from Fig. 5-12 measured on Y123 single crystals have been dashed in. The films tend to have T_c values lower than those found in single crystals, as can be seen for H=0. For the

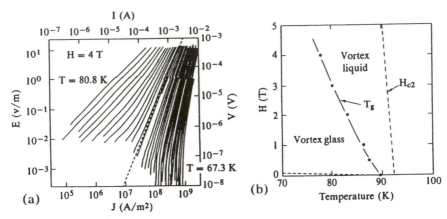

Fig. 6-3 (a) E vs. J for a Y123, c-axis-oriented film in a 4 T field parallel to c. The scale is logarithmic, so it covers many decades of voltage and current. The E vs. J isotherms are taken at 0.3 K intervals, so the dashed line is at $\approx$ 77.4 K (which is T_g). (b) $T_g(H)$ or the equilibrium phase boundary between the vortex-glass phase and the vortex-liquid phase. The dashed line, at very low fields, represents the Meissner phase. H_{c2} vs. T from Fig. 5-12 is added to indicate the boundary of the vortex liquid phase, however; see the text. T_c of the films is usually lower than in single crystals, as can be seen. For details, see R. H. Koch, V. Foglietti, W. J. Gallagher, G. Koren, A. Gupta, and M. P. A. Fisher, Phys. Rev. Lett. **63**, 1511 (1989).

range of fields and temperatures shown, the difference between the T_g and H_{c2} lines is only about 10 K. Thus, the vortex-liquid phase does not exist in a large area of the H-T phase diagram, at least near T_c. However, in other high-T_c materials, 2-Bi(n=2) for example, $T_g(H)$ occurs at much lower fields, resulting in a larger temperature range between T_g and H_{c2}.

6-4f Irreversibility Line — Besides unprecedented transition temperatures, unexpected magnetic behavior was one of the first properties measured in high-T_c superconductors. The magnetic moment (M) vs. temperature is sketched in Fig. 6-4a. After being cooled in zero field (ZFC), M vs. T is measured in a field H ($> H_{c1}$) with increasing temperature. Then M$=-1/4\pi$ or less is obtained at low temperature and M(T) is indicated in the figure. If the sample is cooled in a field (FC), then M(T) is different than the ZFC result, and the FC result is also sketched in Fig. 6-4a. Note; below T_c there is a temperature (T_{irr}) where M(T) of the ZFC and FC results are the same. These measurements can be performed for different magnetic fields and $T_{irr}(H)$ determined. The line of $T_{irr}(H)$, in the H-T plane, is called the **irreversibility line**. Above and to the right of this line,

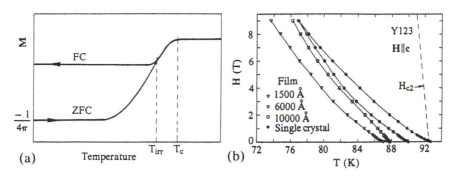

Fig. 6-4 (a) M vs. T, measured in a field H, for a sample cooled in zero field (ZFC), and when the sample is cooled with H applied (FC). (b) T_{irr} of Y123 films of different thicknesses and a single crystal. H_{c2} vs. T from Fig. 5-12 is added to indicate the boundary of the vortex liquid phase. save L. Civale, T. K. Worthington, and A. Gupta, Phys. Rev. B **43**, 5425 (1991).

the sample has reversible magnetic behavior. Below and to the left of this line, the sample has irreversible behavior.

Figure 6-4b shows the measured irreversibility lines for **H** along the **c** axis for a Y123 single crystal and films with different thicknesses. $H_{c2}(T)$, from Fig. 5-12, is added for comparison purposes. The irreversibility lines for the films and crystal are similar except for a shift of T_c. This similarity indicates that the pinning potential-energy minima have about the same depth in Y123 films and crystals. However, films have many more pinning sites than crystals.

The types and depths of pinning sites can be investigated in another manner. Irradiation with neutrons or protons of Y123 single crystals at 77 K can increase J_c (at 1 T) by about 100 times, from 10^3 to 10^5 amp/cm^2. However, the irreversibility line is essentially unshifted by the irradiation. This would indicate that only the density of pinning sites is increased by this type of irradiation. However, irradiation of single-crystal Y123 by heavy ions (e.g., 600 MeV Sn) increases J_c, but also can shift $T_{irr}(H)$ to higher temperatures. This may result from the fact that heavy ions can form a damage track parallel to the **c** axis that may pin vortices in a deeper potential well than before this irradiation.

Comparison of $T_{irr}(H)$ in Fig. 6-4b with $T_g(H)$ in Fig.6-3b, both for **H** along the **c** axis of Y123 films, reveals that they are similar. Since both temperatures are related to vortices overcoming pinning barriers, a close relationship should be expected. Clearly, more experimental and theoretical work is required.

In the preceding discussion, dc results were considered. However, the irreversibility line may be defined as the temperature where vortex relaxation equals the measurement frequency. Thus, T_{irr} could be frequency-dependent.

In the standard Anderson-Kim (thermally activated) flux-creep model, the net potential barrier depends on current density with a linear term (from the Lorentz force) as

$$U_J = U_0[1 - (J/J_{c0})] \qquad (6 - 4c)$$

where J_{c0} is the critical current density in the absence of thermal activation. A vortex-hopping time is taken in the standard Arrhenius manner as

$$t = t_{eff} \exp(U_J/k_BT) \qquad (6 - 4d)$$

where t_{eff} is an effective attempt time for vortex hopping. Combining Eqs. 6-4c and 6-4d, we obtain the useful flux-creep formula,

$$J_c(T,t) = J_{c0}[1 - (k_BT/U_0) \ln(t/t_{eff})] \qquad (6 - 4e)$$

Using the criteria that $J_c(T, t)=0$ at $T=T_{irr}$, then

$$k_BT_{irr}/U_0(T, B) = [\ln(t/t_{eff})]^{-1} = [\ln(\omega_{eff}/\omega)]^{-1} \quad (6 - 4f)$$

where ω_{eff} is the characteristic frequency and, as indicated, the pinning energy may depend on the temperature and magnetic field. Note that in the dc limit ($\omega \to 0$), the Anderson-Kim flux-creep theory predicts that T_{irr} tends to zero. In the vortex-glass model, $T_{irr}(\omega)$ would approach a *finite temperature*, the glass transition temperature T_g, as ω tends to zero.

In conventional superconductors, the vortex-liquid phase is expected to be very narrow, with T_g essentially coinciding with T_{c2}. Enhanced thermal fluctuations, due to higher temperatures, smaller coherence length, and large anisotropy, lead to an observable vortex-liquid region in high-T_c superconductors. In fact, with our present knowledge from the high-T_c materials, preliminary studies of conventional superconductors indicate similar effects but in a much narrower temperature range. The position of the mean-field $H_{c2}(T)$ line is being brought into question: how is it really defined and what is its meaning? The results shown in Fig. 5-2 are determined by magnetization vs. temperature measurements, and H_{c2} seems to be reasonably well-defined from an experimental point of view. However, from a resistance point of view, there is essentially no change in going across $H_{c2}(T)$, so this phase boundary presents interesting questions.

The separation between $H_{c2}(T)$ and $T_{irr}(H)$ in Y123 is relatively small (Fig. 6-4b). However, in 2-Bi(n=2), the other well-studied material, the irreversibility line is found at much lower temperatures. Recent work (Notes) has attributed this effect to the extreme anisotropic nature of this material. The 2-Tl(n) materials seem to behave similarly to 2-Bi(n=2). Thus, so far, only Y123 has a fairly high $T_{irr}(H)$ line, indicating a useful H-T range for applications at 77 K. However, the reversibility region between $T_{irr}(H)$ and H_{c2} can be reduced by pinning sites that have a larger pinning energy U, and there are considerable efforts in this area.

6-5 Applications Introduction

Since the initial discoveries of the families of high-temperature superconductors (1986-1988), there has been a tremendous scientific endeavor to understand their properties, as discussed throughout this book. There also have been very considerable efforts in applications areas.

Of course, there already are applications of conventional superconductors. Thus, the application field is well ground from an engineering point of view, many systems have been studied, and some commercial products exist, such as high-field magnets. Then, for applications of high-T_c superconductors, the questions are: what are the simplifications that high-T_c materials allow over conventional superconductors, and what are the new difficulties to overcome? Naturally, operation at liquid nitrogen may allow for some operations that were impractical with conventional superconductors. At present, it is not yet clear how high-T_c materials will affect technology. However, the amount of effort and the rate of progress is impressive and it appears that commercial applications will follow, and perhaps soon.

Generally speaking, applications of superconductors can be divided into two types. The first are **large-scale applications**, in which large currents and long lengths of superconductors are required in environments where the magnetic field may be several tesla ($1 \text{ T} = 10^4 \text{ Oe}$). Examples include magnets and power transmission lines, transformers, and generators where current densities of at least 10^5 amps/cm^2 are required. For many of the large-scale applications, superconductors are advantageous over normal metals only because of the lower resistance and hence smaller power loss. The present commercial superconductor magnets are typically made of superconducting Nb-Ti (or Nb$_3$Sn) strands imbedded in Cu, with the superconductor composing about one-half of the cross-sectional area. The current densities are up to 10^5 amps/cm^2, with most of it flowing through the superconductor.

The second type of applications are **small-scale applications,** where the currents are much smaller, and the more specialized properties of superconductors tend to be exploited. Examples include detection systems, and analog and digital processing. SQUIDs operating at 77 K could be used in **biomagnetics.** SQUIDs are the most sensitive magnetic field detectors. They are used for measurements of the small magnetic fields from the body or the brain. The SQUIDs could be closer to the body, easier to operate, and more sensitive if only liquid nitrogen is required as a coolant. Such SQUIDs could also be advantageously used to detect small changes in the magnetic field of the earth; for example, in the search for oil, ore, or submarines. Other small scale applications might include current mixers (which produce a difference frequency signal from two carrier frequencies) and computer elements. The latter might include lossless interconnects between the active logic elements or could include the active logic elements themselves.

Certainly, one of the most exciting things about these materials is that superconductivity is achieved at 77 K. Thus, inexpensive and easy-to-use liquid nitrogen can be the coolant. Also, the high-T_c materials are all type II superconductors with enormous H_{c2} values. These two properties make high-T_c materials particularly exciting from an applications point of view.

However, high-T_c metals are not ductile as are Cu, Al, and other normally encountered metals, or as even the Nb-Ti or Nb_3Sn conventional superconductors that can be fabricated into wire and tape. Rather, the high-T_c materials are metals with a brittleness that is closer to that of a ceramic dinner plate. Thus, forming wires, which are needed in many applications, is difficult. Nevertheless, these new materials can be synthesized in thin films, and only time will tell where and how they will make an impact.

6-6 Large-Scale Applications

6-6a Introduction — The costs of energy consumption in the world and the electrical energy in particular are staggering. The elimination of even a small fraction of the resistive load could have a significant impact. Another important impact could arise from the use of high-T_c superconductors in the production of strong magnetic fields above the 2 tesla level. This could eliminate the iron cores in motors, generators, and transformers, resulting in reduced size, weight, and losses from the iron cores.

Eventually, the electric utility industry may become the largest user of superconductors. Certainly, the industry has been researching the use of superconductors for transmission cables, generators, and energy storage.

However, as with any technology, things take time. Uses of high-T_c devices undoubtedly will start with applications that have a small monetary value.

6-6b Wires and Superconducting Magnets — From an applications point of view, superconductors can be characterized by three important parameters, T_c, H_{c2}, and J_c. For the high-T_c superconductors, the former two are very large; it is the latter quantity that is still problematic. Wires of silver-sheathed 2-Bi(n=3) can now be made with properties that are superior to those of conventional superconductors at 4 K to 20 K and at high magnetic fields (Fig. 6-2b). It might be anticipated that laboratory high-field magnetics with this wire will replace those presently sold that use conventional superconductors (Notes).

An important advance also could be made if the irreversibility line in 2-Bi(n=3) could be pushed above 77 K. Since the irreversibility line for Y123 is above 77 K for reasonable field values (Fig. 6-4b), another important advance could be made if it could be learned how to make this material into long wires.

There are many important industrial areas where large-scale superconducting applications could have important implications. This includes energy storage devices (for the electric power industry), electric motor windings (electric motors account for more than 50% of the electrical energy in the western world), electromagnetic pumps, magnets for magnetic separation, magnetic heat pumps, and others. The Notes have some references, but more recent proceedings of applications of superconductivity conferences should be consulted.

6-6c Levitation — There have been many proposals of magnetically levitated trains (**maglevs**). The technology and cost-effectiveness will not be discussed here. However, a simplified introduction into levitation will be presented because of the novelty. The levitation of a superconducting particle above a magnet (or vice versa) is not only an impressive sight. Using high-T_c materials, levitation demonstrations can easily be done in a classroom, since only liquid nitrogen is needed. The experiment is typically done with a magnet under a beaker of liquid nitrogen and by dropping a high-T_c superconductor into the liquid. When cooled below T_c, the superconductor jumps up and hovers above the magnet. If the liquid level and magnetic strength are appropriately arranged, the particle will jump out of the liquid, warm up, and drop back into the liquid, repeating the process.

The repulsion of the particle from the magnet is due to flux exclusion and we examine this effect via a simple calculation. Figure 6-5a shows the

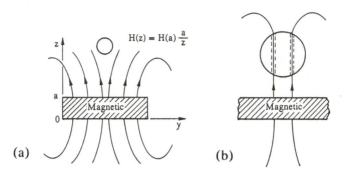

Fig. 6-5 Schematic diagrams to show a superconductor levitating above a magnetic. (a) The z dependence of the magnetic field. (b) Flux lines threading the type II superconductor.

magnetic field, which varies along the z axis, $H(z)$, above the magnetic of height a. We make the simplifying assumption that the field above the magnetic, in the vertical direction, decreases as z^{-1}, $H(z)=H(a)(a/z)$, as indicated in the figure. The magnetic energy density is $H^2/8\pi$, so

$$E_M = \frac{[H(a)]^2}{8\pi} \frac{a^2}{z^2} V \qquad (6-6a)$$

The gravitational potential energy of the superconducting particle is $mgz = \rho Vgz$, where ρ and V are the density and volume of the superconductor. The force on the particle is $F = -\delta E/\delta z$, which for equilibrium is set to zero, yielding

$$z_{equ} = \{[H(a)]^2 a^2/4\pi\rho g\}^{1/3} \qquad (6-6b)$$

For this simple model calculation, we have implicitly assumed $z_{equ} \gg R \gg \lambda$, where R is the radius of the levitating superconducting sphere and λ the magnetic penetration length. Clearly, if R becomes too large, the bottom of the sphere approaches the magnetic, and other forms of $H(z)$ are probably more appropriate. If λ is comparable to R, then the effective superconductor volume is decreased, which would cause the particle to drop. An increase in λ can be accomplished by raising the temperature, which would cause the superconductor to drop. The levitation of high-T_c pressed pellets involves more complicated details than discussed here.

Now that we have seen how flux exclusion produces z-axis levitation, we ask about the x- and y-direction stability. The stability in these directions apparently is caused by flux pinning as indicated in Fig. 6-5b, where details of flux lines threading a type II superconductor are shown. In order for the superconductor to have x- or y-direction motion, pinning forces must be overcome; if they are not, then stability is achieved. Of course, for applications, the particle could be physically constrained in these directions. If the magnetic has low symmetry, then lateral stability is achieved and its position is not unique. In fact, several superconductors can be levitated simultaneously, with all of them behaving as though they were separately "embedded in sand." On the other hand, superconductors with rotational symmetry can have nearly undamped rotation about the symmetry axis.

There are several maglev train test strips and there is talk about a 13-mile commercial line in the Orlando, Florida area and a longer one between Los Angeles and Las Vegas. One proposal is to use an on-board electromagnetic to levitate the train above the laminated iron rail in the guide with ~1 cm air gap. A second proposal is to use superconducting wire coils in the vehicle to produce a magnetic field of the same polarity as coils in the guide's; the repulsive force lifts the vehicle above the track 10-15 cm. The latter version has the advantage that lower tolerances are required in the guide, and since iron is not required for the magnetic, the vehicle could be much lighter.

Progress of maglev trains should be interesting to follow. These trains should be capable of very high speeds so that they could compete with short-hop airplane flights in crowded air corridors.

6-7 Small-Scale Applications

A basic discussion of small-scale applications is difficult because the field is so broad, it is in rapid stages of growth, and it is of a more specialized nature than the other topics covered in Chapter 6. We mention a few generalities, but the reader should consult the Notes for references to books, conferences, and review articles, and one should remain alert for future conference proceedings.

High-T_c thin films have already been fabricated into passive microwave devices such as filters and resonators that could be used in space. Infrared detectors and SQUID magnetometers should also find early commercial uses. A key element for most superconductive devices is the Josephson junction.

Josephson (1962) proposed that in a superconducting-insulator-superconductor (SIS) junction, a *current would be observed at zero voltage*. This is called the **dc Josephson effect**, and this current can be called the **Josephson pair current**, I_0. Josephson further proposed that if a potential,

V, is applied across the junction, then the passage of current would be accompanied by emission of radiation of frequency $h\nu = (2e)(V)$. This is called the **ac Josephson effect**. This frequency is determined by Planck's condition for a particle of charge 2e falling in a potential difference V. This same result can be observed by irradiating a junction with microwaves when a current is flowing in the SIS junction; the current-voltage curve of the junction has voltage steps in integral multiples of $h\nu/2e$, **Shapiro steps**.

The original proposal was for a SIS junction. However, it is generally appreciated that the same ideas apply to "weak links." A **weak link** can be two superconductors separated by a normal metal (SNS), a point contact between two superconductors, or a short constriction in cross section of a superconductor.

For high-T_c materials, the practical challenges of fabricating Josephson junctions are daunting. On each side of the junction, the superconducting order parameter must have nearly its bulk value to within a coherence length of the interface. Working in the **ab** plane, the coherence lengths are only 15Å to 30Å (Section 5-7d). Thus, actual junctions require materials control on the order of a few unit cells, a very difficult task.

Naturally occurring grain boundaries in Y123 films act like Josephson junctions. Thus, useful studies have been carried out on such junctions. By properly orientating and fusing together substrate bicrystals, such junctions can be routinely produced and studied. For example, low-noise SQUIDS have been produced that operate above 77 K. However, extending this technique to complex circuits would seem difficult. Another approach to obtain junctions is to grow the film in a sharply etched step in the substrate. Abrupt orientational changes in the film usually develop grain boundaries. This principle may be more appropriate for yielding complex circuits (Notes).

Progress continues on the development of other building blocks of high-T_c circuit technology; multilayers, for example, which would be required to produce integrated circuits. With the remarkable rate of progress, it is difficult to predict the future, but optimism may be in order.

Problems

1. Persistent current duration — Consider a persistent current circulating in a coil of radius R, which thus has flux lines threading the center of the coil.

The probability per unit time that a fluxoid quantum will leak out of the coil is P=(attempt frequency)(barrier activation factor), where the latter is exp($-\Delta F/k_B T$). Show that

$$\Delta F = (R\xi^2)(H_c^2/8\pi)$$

is a reasonable estimate. For the attempt frequency, use $2\Delta/\hbar$. Estimate P^{-1} and compare it to the age of the universe. What are the temperature dependences of ΔF and P^{-1}? (Hint; see most editions of Kittel.)

2. Critical current of a type I superconductor — Do problem 2 in Chapter 2.

3. Critical current of a type II superconductor — Consider a long, cylindrical superconducting wire. Approximate the GL order parameter $\psi(\mathbf{r})$ by $|\psi|e^{i\phi(\mathbf{r})}$, where $|\psi|$ is constant corresponding to a uniform current flowing through the wire. Write the current from the GL equation for J (Eq. 2-5j), and minimize the GL free energy (Eq. 2-5c) with respect to $|\psi|^2$, neglecting the magnetic energy density term. Thus, show that the expression for J_c in Eq. 6-3a is correct. (Hint; see Tinkham, Section 4-4.)

4. Bardeen–Stephen model — Discuss this model. (See van Duzer and Turner, Section 8.08, and Tinkham, Section 5-5.1.)

5. Read the original **Josephson tunneling** proposal. (See B.D. Josephson, Phys. Letters **1**, 251 (1962).) Is the significance of the proposal apparent? Do you think the author believes it will be experimentally observed? You might also want to read the author's Nobel Prize talk, Rev. Mod. Phys. **46**, 251 (1974).

Bibliography

1a. We list some books and articles on conventional superconductivity (Chapter 2).

D. Shoenberg, "Superconductivity" (Cambridge Univ. Press, 1952).

P. G. De Gennes, "Superconductivity of Metals and Alloys" (Addison-Wesley, 1966 and 1989).

C. G. Kuper, "Introduction to the Theory of Superconductivity" (Clarendon Press, 1968).

A. C. Rose-Innes and E. H. Rhoderick, "Introduction to Superconductivity" (Pergamon Press, 1968).

R. D. Parks, Ed., "Superconductivity" (Dekker, 1969), Vols. **1** and **2**.

L. Solymar, "Superconducting Tunneling and Applications" (Wiley-Interscience, 1972).

M. Tinkham, "Introduction to Superconductivity" (Krieger Pub. Co., 1975 and 1980).

D. H. Douglass, Ed., "Superconductivity in d- and f-Band Metals" (Plenum Press, 1976).

H. W. Weber, Ed., "Anisotropy Effects in Superconductors" (Plenum Press, 1976).

R. W. White and T. H. Geballe, "Long Range Order in Solids" (Academic Press, 1979).

T. Van Duzer and C. W. Turner, "Principles of Superconductive Devices and Circuits" (Elsevier, 1981).

O. Fischer and M. B. Maple, "Superconductivity in Ternary Compounds" (Springer-Verlag, 1982), Vols. 1 and 2.

S. V. Vonsovosky, Y. A. Izyunov, and E. Z. Kurnaev, "Superconductivity of Transition Metals" (Springer-Verlag, 1982).

J. R. Schrieffer, "Theory of Superconductivity" (Addison-Wesley, 1983). This reprinted edition contains the Nobel Lectures of B, C, and S.

K. Kinoshita, Phase Transitions A **23**, 73 (1990).

T. P. Orlando and K. A. Delin, "Foundations of Applied Superconductivity" (Addison-Wesley, 1991).

A **historical perspective** of conventional superconductivity is given in papers by C. J. Gorter and K. Mendelssohn in the proceedings of a 1963 conference on superconductivity, which appears in Rev. Mod. Phys. **36** (1964). The other papers in the proceedings are also worth looking at. H. Thomas, in Bednorz and Müller, Eds. (Bib.), also is excellent.

1b. Some of the key papers of conventional superconductors are:

H. Kamerlingh Onnes, Leiden Comm. 120b, 122b, 124c (1911).

W. Meissner and R. Ochsenfeld, Naturwissenschaften **21**, 787 (1933).

F. and H. London, Proc. Roy. Soc. (London) **A149**, 71 (1935).

A. B. Pippard, Proc. Roy. Soc. (London) **A216**, 547 (1953) and T. E. Faber and A. B. Pippard, ibid **A231**, 336 (1955).

V. L. Ginzburg and L. D. Landau, Zh. Eksperim. i. Teor. Fiz. **20**, 1064 (1950).

F. London, "Superfluids" (Wiley, 1950), Vol. 1.

A. A. Abrikosov, Zh. Eksperim. i Teor. Fiz. **32**, 1442 (1957) [Soviet Phys. JETP **5**, 1174 (1957)].

H. Fröhlich, Phys. Rev. **79**, 845 (1950).

L. N. Cooper, Phys. Rev. **104**, 1189 (1956).

J. Bardeen, L. N. Cooper, and J. R. Schrieffer, Phys. Rev. **108**, 1175 (1957).

I. Giaever, Phys. Rev. Letters **5**, 147 and 464 (1960).

B. D. Josephson, Phys. Letters **1**, 251 (1962).

2. A few books and review articles on high-T_c superconductivity are listed. However, remember that the high-T_c field is rapidly expanding, with many research papers and review articles being published each week. Thus, the literature should be consulted for recent developments. For some of the early reviews of this field, see:

J. C. Phillips, "Physics of High-T_c Superconductors" (Academic Press, 1989).

M. Tinkham and C. J. Lobb, in "Solid State Physics" (Academic Press, 1989), Vol. 42, page 91.

K. C. Hass in "Solid State Physics" (Academic Press, 1989), Vol. 42, page 213.

Many of the highlights are discussed in the "Search and Discovery" section of Physics Today (usually starting on page 17).

"MRS Bulletin" 24 (Number 1), Jan., 1989.

"Physics Today" 44 (Number 6), June, 1991 has some very useful articles, including a discussion between P. W. Anderson and R. Schrieffer that should not be missed.

Some edited books, typically conference proceedings, are listed here.

K. S. Bedell, D. Coffey, D. E. Meltzer, D. Pines, and J. R. Schrieffer, "High Temperature Superconductivity: Proceedings" (Addison-Wesley, 1990).

J. G. Bednorz and K. A. Müller, "Earlier and Recent Aspects of Superconductivity" (Springer-Verlag, 1990).

H. Fukuyama, S. Maekawa, and A. P. Malozemoff, Ed., "Strong Correlation and Superconductivity" (Springer-Verlag, 1989).

D. M. Ginsberg, Ed., "Physical Properties of High-Temperature Superconductors" (World Scientific, 1989 and 1990), Vols. 1 and 2.

S. K. Joshi, C. N. R. Rao, and S. V. Subramanyam, "International Conference on Superconductivity" (World Scientific, 1990).

H. Kamimura and A. Oshiyama, Ed., "Mechanisms of High Temperature Superconductivity" (Springer-Verlag, 1989).

J. W. Lynn, Ed., "High Temperature Superconductivity" (Springer-Verlag, 1990).

S. V. Subramanyam and E. S. Raja Gopal, "High Temperature Superconductors" (Wiley, 1989).

3. Some general solid-state books that may be useful references for particular points are listed here.

N. W. Ashcroft and N. D. Mermin, "Solid State Physics" (Holt, Rinehart, and Winston, 1976).

G. Burns, "Solid State Physics" (Academic Press, 1987).

P. A. Cox, "The Electronic Structure and Chemistry of Solids" (Oxford Univ. Press, 1987).

C. Kittel, "Introduction to Solid State Physics" (John Wiley and Sons, various editions).

Notes for the Chapters

Notes for Chapter 2

General references for this chapter are any of the superconductivity books listed in the Bibliography. In particular, see Tinkham, van Duzer and Turner, and Gladstone, Jensen, and Schrieffer in Parks, Ed. (Bib.), Vol. 2.

T_c values are listed in Table 2-1. However, more complete complications of T_c and $H_c(0)$ can be found in B. W. Roberts, NBS Technical Note 983 (1978); S. V. Vomsovsky, Y. A. Izyumov, and E. Z. Kurmaey, "Superconductivity of Transition Metals" (Springer-Verlag, 1982); CRC Handbook of Chemistry and Physics (1983-1984); Kinoshita (Bib.).

For the two-fluid model, see C. J. Gorter and H. G. B. Casimer, Phys. Z. **35**, 963 (1934) and Z. Tech. Phys. **15**, 539 (1934).

Chambers equation (Eq. 2-3a) is discussed in J. M. Ziman, "Principles of the Theory of Solid State Physics" (Cambridge Univ. Press, 1972) pp. 283 and 402. For **nonlocal electrodynamics**, see A. B. Pippard, Proc. Roy. Soc. (London) **A216**, 547 (1953) and T. E. Faber and A. B. Pippard, ibid **A231**, 336 (1955). Also see D. C. Mattis and J. Bardeen, Phys. Rev. **111**, 412 (1958).

London equations are discussed in F. London, "Superfluids" (Wiley, 1950), Vol. 1, and most of the superconductivity texts in the Bibliography.

Ginzburg-Landau theory is discussed in many books. See Tinkham, and van Duzer and Turner (Bib.), for a more complete discussion than presented here.

BCS theory is discussed in the elementary and intermediate SSP books listed in the Bibliography. The original paper (well worth reading) is J. Bardeen, L. N. Cooper, and J. R. Schrieffer, Phys. Rev. **108**, 1175 (1957).

P. W. Anderson and R. Schrieffer discuss many aspects of the theory of superconductivity in Physics Today **44** (Number 6), June, 1991. This discussion should not be missed.

Specific heat results can be found in many places. Some recent complications are in Lynn and Kinoshita (Bib.). References to the earlier literature are cited in these compilations. Even in the conventional superconductors, there are many deviations from the predicted specific heats at low temperatures. For a discussion (and references) of these results, see the article by Meservey and Schwartz, as well as one by Gladstone, Jensen, and Schrieffer, both in Parks, Ed. Also see L. L. Daema and A. W. Overhauser, Phys. Rev. B **39**, 6431 (1989).

Coherence effects, as predicted by BCS, are discussed in Tinkham (Section 2-9) and Schrieffer (Chapter 3), where references to experiments can be found. **Nuclear magnetic resonance** in superconductors is reviewed by D. E. MacLaughin, in "Solid State Physics" **31**, 1 (1976) and C. H. Pennington and C. P. Slichter, in Ginsberg, Ed., Vol 2, p. 269 (Bib.). Discussion of NMR coherence effect observations in terms of an anisotropic superconducting gap and strong-coupled BCS are in Y. Masuda, Phys. Rev. **126**, 1271 (1962) and M. Fibich, Phys. Rev. Letters **14**, 561 (1965). Also see the Notes for Chapter 5.

Strong-coupled BCS results are discussed (and referenced) in the articles by Meservey and Schwartz, McMillian and Rowell, and Scalapino, all in Parks, Ed. (Bib.). Also see P. B. Allen and B. Mitrovic, "Solid State Physics" (Academic Press, 1982), Vol. 37, p. 1, with references on **maximum** T_c **values**, contained in Section 21. W. L. McMillan's readable, original paper is Phys. Rev. **167**, 331 (1968), where references to Migdal, Eliashberg, and Nambu can be found. Also see P. B. Allen and R. C. Dynes, Phys. Rev. B **12**, 905 (1975), where a table of electron-phonon coupling parameters is given for elements and alloys that have large λ_{ep} values. Phillips (Bib.) discusses systematics of λ_{ep} in some of the conventional superconductor-alloy systems.

Tunneling, in general, is treated in Chapter 2 of Tinkham, as well as in other books. Tunneling in strong-coupled superconductors (Pb in particular) is discussed in the article by W. L. McMillan and J. M. Rowell in Parks, Ed., Vol. 1. Also see E. Burstein and S. Lindqvist, Eds., "Tunneling Phenomena in Solids" (Plenum, 1969) and E. L. Wolf, "Principles of Tunneling Spectroscopy" (Oxford, 1985).

Magnetic superconductors are covered in M. B. Maple and O. Fischer, "Superconductivity in Ternary Compounds" (Springer-Verlag, 1983), Vol. 2. See D. H. Douglass, Ed., "Superconductivity in d- and f-band Metals", (Plenum Press, 1976). Also see Kinoshita (Bib.), Sections 3.4 and 3.5, and L. N. Bulaevskii, A. I. Buzdin, M. L. Kulic, and S. V. Panjukov, Adv. in Phys. **34**, 175 (1985).

Heavy-electron materials are reviewed in G. R. Stewart, Rev. Mod. Phys. **56**, 755 (1984); P. A. Lee, T. M. Rice, J. W. Serene, L. J. Sham, and J. W. Wilkins, Comments Cond. Mat. Phys. **12**, 99 (1986); and P. Fulde, J. Keller, and G. Zwicknagl, "Solid State Physics" (Academic Press, 1988), Vol. 41, p. 1.

Anisotropic energy gaps in lead are calculated by A. Bennett, Phys. Rev. A **140**, 1902 (1965), and many references are included there. See H. W. Weber, Ed., "Anisotropic effects in Superconductors" (Plenum Press, 1977). See Allen and Mitrovic (Section 16), mentioned above. Also see Meservey and Schwartz, in Parks, Ed., Vol. 1.

Organic superconductors are discussed by T. Ishiguro and K. Yamaji, "Organic Superconductors" (Springer-Verlag, 1990); L. N. Bulaevskii, Adv. Phys. **37**, 443 (1988). D. Jéome, in Bednorz and Müller, Eds. (Bib.). U. Geiser, et al., Physica C **174**, 475 (1991).

^{3}He is reviewed in papers by A. J. Leggett and J. C. Wheatley in Also see the two review articles by P. W. Anderson and W. F. Brinkman, and by D. M. Lee and R. C. Richardson, in K. H. Bennemann and J. B. Ketterson, Eds., "The Physics of Liquid and Solid Helium&eqo,. Part 2 (Wiley, 1978), Chapters 3 and 4.

Proximity effects refer to effects that occur when there is excellent contact between a normal metal (N) and a superconductor (S). The Cooper pairs can leak from S to $\sim 10^3$Å into N, which means that thin-metal films can be used. The effects depend on whether the mean free path in N is large or small compared to the coherence length in S (clean and dirty limit). See G. Deutscher and P. G. De Gennes, in Parks, Ed., Vol. 2.

Historical point of view is taken in an excellent review by H. Thomas, in Bednorz and Müller, Eds. (Bib.).

Metallic hydrogen is considered in T. W. Barbee, A. Garcia, and M. L. Cohn, Nature **340**, 369 (1989), where μ^*, λ_{ep} and Θ_D are calculated.

Notes for Chapter 3

Reviews of the structural aspects, including modulation structures and other distortions, of high-temperature superconductors can be found in the following:

G. Burns and A. M. Glazer, "Space Groups for Solid State Scientists" (Academic Press, 1990), Chapter 9. This book also contains a discussion of space groups applied to other structures.

R. Byers and T. M. Shaw in "Solid State Physics" (Academic Press, 1989), Vol. 42, page 135.

K. Yvon and M. Francois, Z. Phys. B **76**, 413 (1989).

H. Müller-Buschbaum, Angew. Chem Int. Ed. Engl. **28**, 1472 (1989).

R. M. Hazen, in Ginsberg, Ed., Vol. 2 (Bib.).

See the articles by J. D. Jorgensen and A. W. Sleight in Physics Today **44**, June, 1991.

Crystal structures and related aspects are covered in the above publications, where many references can be found. Here, we list just a few papers covering recent aspects discussed in this chapter:

P. Haldar, K. Chen, B. Maheswaran, A. Roig-Janicki, N. K. Jaggi, R. S. Markiewicz, and B. C. Giessen, Science 241, 1198 (1988) discusses 1-Tl(n=4).

V. I. Simonov, L. A. Muradyan, R. A. Tamazyan, V. V. Osiko, V. M. Tatarintsev, and K. Gamayumov, Physica C **169**, 123 (1990) discusses the ordering of Sr in single-crystal La(n=1).

H. Shibata, K. Kinoshita, and T. Yamada, Physica C **170**, 411 (1990) discusses the doping of La(n=1) with Na and K.

R. J. Cava, B. Batlogg, R. B. van Dover, J. J. Krajewski, J. V. Waszczak, R. M. Fleming, W. F. Peck Jr., L. W. Rupp Jr., P. Marsh, A. C. W. P. James, and L. F. Schneemeyer, Nature **345**, 602 (1990) discusses La(n=2).

Materials with the **n = infinity structure** have now been reported to be superconducting with $T_c \sim 90$ K. The material has the approximate formula $(Sr_{0.7}Ba_{0.3})CuO_2$ and it is formed under high-pressure conditions. See M. Takano, M. Azuma, Z. Hiroi, Y. Bando, and Y. Takeda, Physica C, to be published in 1991.

A **frozen phonon calculation** of the tilting mode in La_2CuO_4 is in R. E. Cohn, W. E. Pickett, and H. Krakauer, Phys. Rev. Lett. **62**, 831 (1989), where references to the neutron measurements can be found.

Doping with excess oxygen is discussed in a neutron diffraction study of 2-Tl(n=1), where T_c can be varied from 0 to 73 K by changing the oxygen content, which changes the (hole) carrier content. Excess oxygen is at an interstitial site between the double Tl-O planes. See Y. Shimakawa, Y. Kubo, T. Manako, H. Igarashi, F. Izumi, and H. Asano, Phys. Rev. B **42**, 10165 (1990).

Joint x-ray and neutron refinement and **pulsed-neutron diffraction** experiments have been reported by A. Williams, G. H. Kwei, R. B. Von Dreele, A. C. Larson, I. D. Raistrick, and D. L. Bish, Phys. Rev. B **37**, 7960 (1988) and G. H. Kwei, A. C. Larson, W. L. Hults, and J. L. Smith, Physica C **169**, 217 (1990). References to many earlier diffraction papers can be found there.

Superstructure diffraction in Y123, oxygen ordering in Y123, vacancy ordering and **Ortho II** are discussed in R. Beyers, B. T. Ahn, G. Gorman, V. Y. Lee, S. S. P. Parkin, M. L. Ramirez, K. P. Roche, J. E. Vazquez, T. M. Gur, and R. A. Huggins, Nature 340, 619 (1989); C. Chaillot, M. A. Alario-Franco, J. J. Copponi, J. Chenavais, P. Strobel, and M. Marezio, Solid State Commun. **65**, 283 (1988); ibid, Phys. Rev. B 36, 7118 (1987); H. F. Poulsen et al., Nature 349, 594 (1991); J. D. Jorgensen, ibid, 565 (1991). Also, for a brief discussion and a list of some original papers, see B. W. Veal, A. P. Paulikas, H. You, H. Shi, Y. Fong, and J. W. Downey, Phys. Rev. B **42**, 6305 (1990). References to calculations of the oxygen-vacancy orderings are also given in this paper.

Notes for Chapter 4

Molecular orbitals in relation to bonds are discussed in R. Hoffmann, "Solids and Surfaces" (VCH Publishers, 1988); Angewandte Chemie, International Ed., **26**, 846 (1987);

and Rev. Mod. Phys. **60**, 846 (1987). Secular determinants are discussed in many quantum texts, or see G. Burns, "Group Theory" (Academic Press, 1977), Chapter 10.

Ziman, "Electrons and Phonons" (Oxford Univ. Press, 1960), Chapter 9, has a quantitative discussion of the temperature dependence of the **electrical resistivity** of normal metals. A discussion of other transport properties can also be found there.

The **Grüneisen-Bloch formulae**, for the temperature dependence of the resistivity, is quantitatively discussed in J. M. Ziman, "Electrons and Phonons."

Resistivity of organic layered superconductors is discussed in references in the Notes to Chapter 2.

Linear resistivity behavior is theoretically discussed by C. P. Enz, Z. Phys. B **80**, 317 (1990). Some recent and interesting experimental data is in T. Ito, H. Takagi, S. Ishibashi, and S. Uchida, Nature **350**, 596 (1991), where reproducible metallic resistivity along the **c** axis is found. It is also found for a certain range of Sr doping in La($n=1$). However, for the other high-T_C materials, nonmetallic conductivity is found along the **c** axis, as in Fig. 4-3c.

The absence of **EPR signals** in high-T_C materials is discussed by F. Mehran and P.W. Anderson, Solid State Commun. **71**, 29 (1989). R. Janes, et al., Solid State Commun. **79**, 241 (1991).

Hole measurement experiments, and their relation to band calculations, is discussed by M. W. Shafer and T. Penney, Euro. J. Solid State Inorg. Chem. **27**, 191 (1990). The chemical (redox titration) method to determine the hole concentration is discussed in this review article. Also see the reference in the paper quoted in Fig. 4-4.

Fermi-liquid theory was first discussed by L. D. Landau, Sov. Phys. JEPT **3**, 920 (1957); **5**, 101 (1957); and **8**, 70 (1959). For an elementary discussion, see D. Pines and P. Nozieres, "The Theory of Quantum Liquids" I (Benjamin, 1966). Also see Ashcroft and Mermin, Chapter 17 (Bib.).

Normal state properties, from a Fermi-liquid theory point of view are review by: K. Levin, J. H. Kim, J. P. Lu, and Q. Si, Physica C **175**, 449 (1991); P. A. Lee in Bodell, et al., Eds. (Bib.).

Resonating-valance-band state was first investigated by P. W. Anderson, Mat. Res. Bull. **8**, 153 (1973), and see Science **235**, 1196 (1987). Also see P. W. Anderson and Z. Zou, Phys. Rev. Lett. **60**, 132 (1988), and N. Nagaosa and P. A. Lee, ibid, **64**, 2450 (1990). What might be called the short-range RVB state is discussed by S. A. Kivelson, D. S. Rokhsar, and J. P. Sethna, Phys. Rev. B **35**, 8865 (1987). The RVB point of view is forcefully and extensively discussed in a forthcoming book by P. W. Anderson and colleagues. We refer the interested reader to this book. P. W. Anderson in Bodell, et al., Eds. (Bib.).

Hubbard model — see J. Hubbard, Proc. Roy. Soc. **A276**, 238 (1963); **A277**, 237 (1964); **A281**, 401 (1964). The model is also discussed in many books; for example, see Cox (Bib.).

Anyons — Elementary discussions of anyons can be found in several places; C. S. Canright and M. D. Johnson, Comments Cond. Mat. Phys. **15**, 77 (1990); F. Wilczek, Scientific America, May, 1991, page 58.

Two-dimensional effects — Dimensional effects, lower than three, have been discussed as possible causes for higher T_C values, even for the conventional A15 compounds. For the high-T_C materials, many people have considered how the two-dimensional aspects of the structure can affect the high values of T_C as well as other normal-state and superconducting properties. We list just a few references here and suggest that the current literature be consulted: J. Friedel, J. Phys. Condens. Matter **1**, 7757 (1989); J. E. Hirsch and D. J. Scalapino, Phys. Rev. Lett. **56**, 2732 (1986); J. Labbe and J. Bok, Europhys. Lett. **3**, 1225 (1987); R. S.

Markiewicz, J. Phys. Condens. Matter **2**, 665 (1990); and Physica C **177**, 171 (1991); C. C. Tsuei, D. M. Newns, C. C. Chi, and P. C. Pattnaik, Phys. Rev. Lett. **65**, 2724 (1990); D. M. Newns, C. C. Tsuei, P. C. Pattnaik, and C. Kane, Comments Cond. Mat. Phys., to be published.

Linear to quadratic resistivity behavior — Some useful temperature-dependent resistivity $\rho(T)$ in 2-Tl(n=1) has appeared. For the hole doping that gives the highest T_c values, $\rho(T)$ is linear in the normal state. As the hole doping is increased, so that the material is a better metal and T_c decreases, $\rho(T)$, in the normal state, becomes more quadratic. See Y. Kubo, Y. Shimakawa, T. Manako, and H. Igarashi, Phys. Rev. B **43**, 7875 (1991).

First Brillouin zones for essentially all of the 14 Bravais lattices with the special points and lines, labeled in the conventional manner, are given by G. F. Koster, Solid State Physics (Academic Press, 1957), Vol. 5, p. 173. For related pictures of all 24 of these zones, see Burns and Glazer, Appendix 3; the **asymmetric unit** or **irreducible wedge** is also discussed in Section 7-3 in this book. It is the piece of the Brillouin zone that is delineated by the special lines in Fig. 4-7. Occasionally, calculations of bands are presented with labels other than those of Koster's; be careful.

For an elementary discussion of **photoemission spectroscopy**, or **PES**, see any elementary solid-state physics text. **Angle-resolved PES** is reviewed by F. J. Himpsel, Adv. in Phys. **32**, 1 (1983).

Magnetism and superconductivity is discussed in Sections 2-9a and 5-7. For their coexistence in Nd(n=1)-type materials, see Y. Dalichaouch, B. W. Lee, C. L. Seaman, J. T. Markert, and M. B. Maple, Phys. Rev. Lett. **64**, 599 (1990).

Notes for Chapter 5

T_c **vs. doping** is discussed in H. Takagi, T. Ido, S. Ishibashi, and S. Uchida, Phys. Rev. B **40**, 2254 (1989); M. W. Shafer and T. Penney, Eur. J. Solid State Inorg. Chem. **27**, 191 (1990).

Paired electrons — See the reference in Fig. 5-2a for fluxoid quantization in a ring, and C. E. Gough in Bednorz and Müller, Eds. (Bib.). **Shapiro steps (ac Josephson effect)** also indicate that the superconducting carriers are paired electrons. See J. Niemeyer, M. R. Dietrich, and C. Politis, Z. Phys. B **67**, 155 (1987).

Singlet spin, s-state paired electrons? — The Josephson tunneling experiment by Niemeyer et al. is one such experiment; see O. S. Akhryamov, Sov. Phys. JEPT Lett. **3**, 183 (1966). Also see the references in Fig. 5-2c. W. A. Little, Science **242**, 1390 (1988) is a short, easily readable review. J. F. Annett, N. Goldenfeld, and S. R. Renn, in Ginsberg, Ed. (Bib.) has a detailed review for this and related subjects. In this latter review, the richness and subtlety of the subject can be appreciated with the authors pointing out other possibilities. Also, **Andreev scattering** experiments are discussed, which also lead to the conclusion that high-T_c superconductivity is associated with **paired electrons**. Also see J. F. Annett, Adv. in Phys. **39**, 83 (1990).

Anisotropic Ginzburg–Landau theory is discussed in R. A. Klemm and J. R. Clem, Phys. Rev. B **21**, 1868 (1980) and V. G. Kogan, Phys. Rev. B **24**, 1572 (1981).

Tunneling is reviewed in J.R. Kirtley, Int. J. of Mod. Phys. **B 4** , 201 (1990). The anisotropy and temperature dependence in 2-Bi(n=2) is discussed in M. Boekholt, M. Hoffmann, and G. Güntherodt, Physica C **175**, 127 (1991).

Specific heat results are controversial and, at times, inconsistent with each other. This may be due to impurities in the (usually) powdered samples or it may be more fundamental. We list a few recent references that could be used as a guide to the literature: N. E. Phillips, R. A. Fisher, and J. E. Gordon, Progress in Low-Temperature Physics 13, to be published, and earlier reviews by these authors, J. of Superconductivity **1**, 231 (1988) and Physica C **153 - 155**, 1092 (1988). Also see A. Junod, in Ginsberg, Ed. **2**, 13 (Bib.), J. E. Crow and N. P. Ong, in Lynn, Ed. p. 302 (Bib.), and N. E. Phillips, R. A. Fisher, J. P. Gordon, S. Kim, A. M. Stacy, M. K. Crawford, and E. M. McCarron III, Phys. Rev. Lett. **65**, 357 (1990).

. **Debye temperatures** for the high-T_c superconductors are listed in the second reference given above. For a more general discussion of the temperature dependence of Θ_D, see Burns (Bib.) particularly Fig. 11-6 and F. W. de Wette, A. D. Kulkarni, J. Prade, U. Schröder, and W. Kress, Phys. Rev. B **42**, 6707 (1990).

Results for the specific heat in $(Ba_{0.6}K_{0.4})BiO_3$ and an organic superconductor can be found in S. J. Collocott, N. Savvides, and E. R. Vance, Phys. Rev. B **42**, 4794 (1990) and J. E. Graebner, R. C. Haddon, S. V. Chichester, and S. H. Glarum, Phys. Rev. B **41**, 4808 (1990).

Specific heat of heavy-electron systems is reviewed by G. R. Stewart, Rev. Mod. Phys. **56**, 755 (1984), who discusses the **Wilson ratio** in Section III-6. The latter is obtained by renormalization group methods; see K. G. Wilson, Rev. Mod. Phys. **47**, 773 (1975).

Double-well or anharmonic potentials for the atoms, and, in particular, the apical-oxygen atoms, in Y123 have been experimentally sought by many works. Claims for both their absence and existence have been published. We list some of the references: (a) A. Williams, G. H. Kwei, R. B. Von Dreele, A. C. Larson, I. D. Raistrick, and D. L. Bish, Phys. Rev. B **37**, 7960 (1988). (b) G. H. Kwei, A. C. Larson, W. L. Hults, and J. L. Smith, Physica C **169**, 217 (1990). (c) J. Mustre de Leon, S. D. Conradson, I. Batistic, and A. R. Bishop, Phys. Rev. Lett. **65**, 1675 (1990). (d) D. Yoshioka, J. Phys. Soc. Japan **59**, 2627 (1990). (e) K. A. Müller, **80**, 193 (1990).

There have been many theoretical studies of the effect of double-well and anharmonic potentials; some of these are listed here: J. R. Hardy and J. W. Flocken, Phys. Rev. Lett. **60**, 2191 (1988); J. Kasperczyk and H. Büttner, Solid State Commun. **75**, 105 (1990). Phillips (Bib.), Chapter 4, Sections 8 and 13, and Phys. Rev. B **42**, 8623 (1990), with references therein, makes strong arguments that the electron-phonon coupling parameter is enhanced due to defects and the quasi-two-dimensionality of the structure.

Electronic Raman scattering associated with the gap is reviewed in S. L. Cooper and M. V. Klein, Comments Cond. Mat. Phys. **15**, 99 (1990) and F. Shakey, M. V. Klein, J. P. Rice, and D. M. Ginsberg, Phys. Rev. B **42**, 2643 (1990). The Nb_3Sn and V_3Si is in R. Hockl, R. Kaiser, and S. Schicktanz, J. Phys. C **16**, 1729 (1983) and S. B. Dierker, M. V. Klein, G. W. Welb, and Z. Fisk, Phys. Rev. Lett. **50**, 853 (1983).

Sharp line Raman scattering associated with the gap is discussed in B. Friedl, C. Thomsen, E. Schönherr, and M. Cardona, Solid State Commun.**76**, 1107 (1990) and references therein.

Infrared measurements have been carried out on high-T_c superconductors to determine the infrared active phonons. See the review by T. Timusk and D. B. Tanner, in Ginsberg, Ed. (Bib.). For some temperature-dependent results, see L. Genzel, A. Wittlin, M. Bauer, M. Cardona, E. Schönherr, and A. Simon, Phys. Rev. B **40**, 2170 (1989).

References to IR measurements to determine the superconducting gap are listed in the captions of Figs. 5-7a to 5-7d. It is felt by many workers that the superconducting gap is not observed in the IR spectra.

Raman measurements are reviewed by R. Feile, Physica C **159**, 1 (1989). The atomic motion for most of these modes is understood (G. Burns, M. K. Crawford, F. H. Dacol) and N. Herron, Physica C **170**, 80 (1990)), and lattice dynamic calculations have been performed (A. D. Kulkarni, F. W. de Witte, J. Prade, U. Schröder, and W. Kress, Phys. Rev. B **41**, 6409 (1990)). Small changes in the frequency of some Raman modes in Y123 and 2-Bi(n=2) have been reported by R. M. MacFarlane, H. J. Rosen, and H. Seki, Solid State Commun.63, 831 (1987) and G. Burns, G. V. Chandrashekhar, F. H. Dacol, and P. Strobel, Phys. Rev. B **39**, 775 (1989).

Coherence effects as measured by **nuclear magnetic resonance** have been theoretically calculated using strong-coupled BCS and shown to be sharply reduced with increased electron-phonon coupling. See P. Allen and D. Rainer, Nature **349**, 396 (1991) and R. Akis and J. P. Carbotte, Solid State Commun. **78**, 393 (1991). For recent measurements see J. C. Jol, et al., Physica C **175**, 12 (1991). Also see the Notes for Chapter 2 for review articles.

Pair breaking in superconductors is discussed by P. Fulde and G. Zwicknagl in Bednorz and Müller, Eds. (Bib.).

Muon spin rotation in high-T_c materials is discussed by J. Keller in Bednorz and Müller, Eds. (Bib.).

Ion channeling results are discussed in R. P. Sharma, L. E. Rehn, P. M. Baldo, and J. Z. Liu, Phys. Rev. Lett. **62**, 2869 (1989); T. Haga, K. Yamaya, Y. Abe, Y. Tajima, and Y. Hidaka, Phys. Rev. B **41**, 826 (1990); L. E. Rehn, R. P. Sharma, P. M. Baldo, Y. C. Chang, and P. Z. Jiang, Phys. Rev. B **42**, 4175 (1990). For **neutron resonance absorption spectroscopy** (NRAS) and related results, see: H. A. Mook, M. Mostoller, J. A. Harvey, N. W. Hill, B. C. Chakoumakos, and B. C. Sales, Phys. Rev. Lett. **65**l, 2712 (1990); B. H. Toby, T. Egami, J. D. Jorgensen, and M. A. Subramanian, Phys. Rev. Lett. **64**, 2414 (1990). **Neutron diffraction and ion-channeling resutls** are discussed in R. P. Sharma, et al., Physica C **174**, 409 (1991).

Inelastic neutron scattering from powders yields phonon density of states, and results are discussed in the references in Figs. 5-9a and 5-9b. For single crystal-results on La_2NiO_4 and La(n=1), see L. Pintschovius, J. M. Bassat, P. Odier, F. Gervais, G. Chevrier, W. Reichardt, and F. Gompf, Phys. Rev. B **40**, 2229 (1989), and L. Pintschovis, in "Phonon Physics," Proceedings of the Third Conference on Phonon Physics, edited by S. Hunklinger, W. Ludwig, and G. Weiss (World Scientific, Singapore, 1989), p. 217. For preliminary single-crystal Y123 results see W. Reichardt, N. Pyka, L. Pintschovius, B. Hennion, and G. Collin, Physica C **162-164**, 464 (1989).

Tunneling measurements in 2-Bi(n=2) are discussed by N. Miyakawa, D. Shimada, T. Kido, and N. Touda, J. Phys. Soc. Jpn. **58**, 383 (1989). Phonon density of states obtained by neutron diffraction is in Fig. 5-9b. Results for Nd(n=1), discussed in Section 5-6e, are in Q. Huang, J. F. Zasadzinski, N. Tralshawala, K. E. Gray, D. G. Hinks, J. L. Peng, and R. L. Greene, Nature **347**, 369 (1990). IR and Raman results for the latter material are in M. K. Crawford, G. Burns, G. V. Chandrashekhar, F. H. Dacol, W. E. Farneth, E. M. McCarron, and R. J. Smalley, Phys. Rev. B **41**, 8933 (1990). Clearly, neutron measurements will be done soon.

Electron-phonon coupling parameter calculations in Y123 are discussed in: R. E. Cohen, W. E. Pickett, and H. Krakauer, Phys. Rev. Lett. **64**, 2575 (1990); R. Zeyher, Z. Phys. B **80**, 187 (1990); C. O. Rodriguez, A. I. Liechtenstein, I. I. Mazin, O. Jepsen, O. K. Andersen, and M. Methfessel, Phys. Rev. B **42**, 2692 (1990). Some experimental values of $\lambda=0.9$ in Y123 have been reported by S. D. Brorson, A. Kazeroonian, D. W. Face, T. K. Cheng, G. L. Doll,

M. S. Dresselhaus, G. Dresselhaus, E. P. Ippen, T. Venkatesan, X. D. Wu, and A. Inam, Solid State Commun. **74**, 1305 (1990).

Phonons alone (Section 5-6h) is discussed in Y. Shiina and Y. O. Nakamura, Solid State Commun. **76**, 1189 (1990).

H_{c1} **and** λ references are given in Fig. 5-11. Also see D. R. Harshman, et al., Phys. Rev. B **36**, 2386 (1987), where a comparison to **heavy-electron superconductors** λ values is given. Measurements of H_{c1} in Y124 are given by J. C. Martinez, J. J. Preuean, J. Karpinski, E. Kaldis, and P. Bordet, Solid State Commun. **75**, 315 (1990). Also see the Note on Bitter patterns.

Dimensional crossover is discussed in the reference in Fig. 5-12. Also see D. E. Farrell, J. P. Rice, D. M. Ginsberg, and J. Z. Liu, Phys. Rev. Lett. **64**, 1573 (1990).

Torque magnetometry is discussed in V. G. Kogan, Phys. Rev. B **38**, 7049 (1988), and recent results are reported by D. E. Farrell, R. G. Beck, M. F. Booth, C. J. Allen, E. D. Bukowaki, and D. M. Ginsberg, Phys. Rev. B **42**, 6758 (1990).

Bitter pattern results of the anisotropy of the penetration depth in Y123 are in G. J. Dolan, F. Holtzberg, C. Feild, and T. R. Dinger, Phys. Rev. Lett. **62**, 2184 (1989), where references to other work can be found.

For a recent review, of high-T_C superconducting properties in general, see B. Batlogg in Physics Today **44**, (1991).

Notes for Chapter 6

Texts that cover many of the topics discussed in this chapter are Tinkham (Section 4-4 and Chapter 5) and van Duzen and Turner (Chapter 8).

Superlattices See Section 6-3b and for recent work: P. Svedlindh, et al., Physica C **176**, 336 (1991). M. Affronte, et al., Phys. Rev. B **43**, 11484 (1991).

Intercalation is discussed in X. D. Xiang, et al., Phys. Rev. B **43**, 11496 (1991).

Magnetic wires of conventional superconductors were originally restricted to ductile alloys, principally Nb-Ti. However, since the development of the "bronze process," brittle Nb_3Sn is widely used in superconducting magnetics up to 20 T. In this process, Nb_3Sn is produced in filamentary conductors at the interface of Nb and Cu-Sn composites by appropriate heat treatment. Wires of 2-Bi(n=3) were discussed in Section 6-6h. Also see S. Jin, et al., Physica C **177**, 189 (1991).

Depairing current is discussed in E. J. Nicol and J. P. Carbotte, Phys. Rev. B **43**, 10210 (1991), where references to the original papers can be found.

Flux creep and **resistance in type II superconductors** are theoretically discussed by P. W. Anderson and Y. B. Kim, Rev. Mod. Phys. **36**, 39 (1964). Also see M. R. Beasly, R. Fabusch, and W. W. Webb, Phys. Rev. **81**, 682 (1969) and A. M. Campbell and J. E. Evetts, Adv. Phys. **21**, 199 (1972). For extensions of the Anderson-Kim model to high-T_C materials, see Y. Yeshurun and A. P. Malozemoff, Phys. Rev. Letters **60**, 2202 (1988) and M. Tinkham, Phys. Rev. Letters **61**, 1658 (1988).

Vortex–glass to vortex–liquid phase transition model is discussed by M. P. A. Fisher, Phys. Rev. Letters **62**, 1415 (1989), and see the more recent work of D. S. Fisher, et al., Phys. Rev. B **43**, 130 (1991). The experimental work (Fig. 6-3) is not uncontroversial. See the comment by S. N. Coppersmith, M. Inui, and P. B. Littlewod, Phys. Rev. Letters **64**, 2585 (1990) and the reply by R. H. Koch, et al., on the next page. However, more recent work goes to lower currents and tends to agree with the results of Koch et al.; L. Gammel et al., Phys.

Rev. Letters **66**, 953 (1991). Results related to those in Fig. 6-3, but in Y123 ceramics, are in T. K. Worthington, E. Olsson, C. S. Nichols, T. M. Shaw, and D. R. Clark, Phys. Rev. B **43**, 10538 (1991).

Papers describing **applications** of superconductors, and high-T_c ones in particular, are spread widely throughout the literature. For a basic point of view, one could start with Tinkham, van Duzer, Turner, and Bedard's article in Lynn, Ed., all in the Bibliography, L. Solymar, "Superconductivity Tunnelling and Applications" (Wiley-Interscience, 1972), and T. P. Orlando and K. A. Delin, "Foundations of Applied Superconductivity" (Addison-Wesley, 1991) is an excellent recent text. For conference proceedings, see "Superconductivity and its Applications," Y. H. Kao, P. Coppers, and H. S. Kwok, Eds. (American Institute of Physics, 1991). **Applied superconductivity conference** (1990) proceedings are published in IEEE Trans. on Magnetics **27**, 814-3405 (1991). This semiannual conference covers a broad range of uses of high-T_c and conventional superconductors in devices.

Applications are also discussed in the articles by D. Larbalestier and R. Simon in Physics Today **44**, (1991).

Levitation is discussed by E. H. Brandt, J. Phys. **58**, 43 (1990), and references to other works can be found in this paper. For a broader discussion of levitation by superconductors as well as other methods, see E. H. Brandt, Science **243**, 349 (1989) and references therein.

Index